Shuriken
An
Illustrated
Guide

Fujita Seiko

Translated by Eric Shahan

Notice:
This book is being translated for historical research purposes only. The translator does not make any representation, warranty or guarantee that the techniques described or illustrated in this book are safe or effective in any self-defense situation. You may be injured if you apply or train Shuriken may or may not be legal in your country, area or state.

Translator's Introduction
This book was originally printed horizontally, and this translation is printed vertically.

図解 手裏剣術 藤田西湖著

ZUKAI
SHURIKEN
JUTSU
BY
FUJITA
SEIKO

Table of Contents

Duel between Shuriken and a folding fan
Fujita Seiko copy of an illustration from Hokusai's *Manga Guide to Martial Arts*, published in 1814

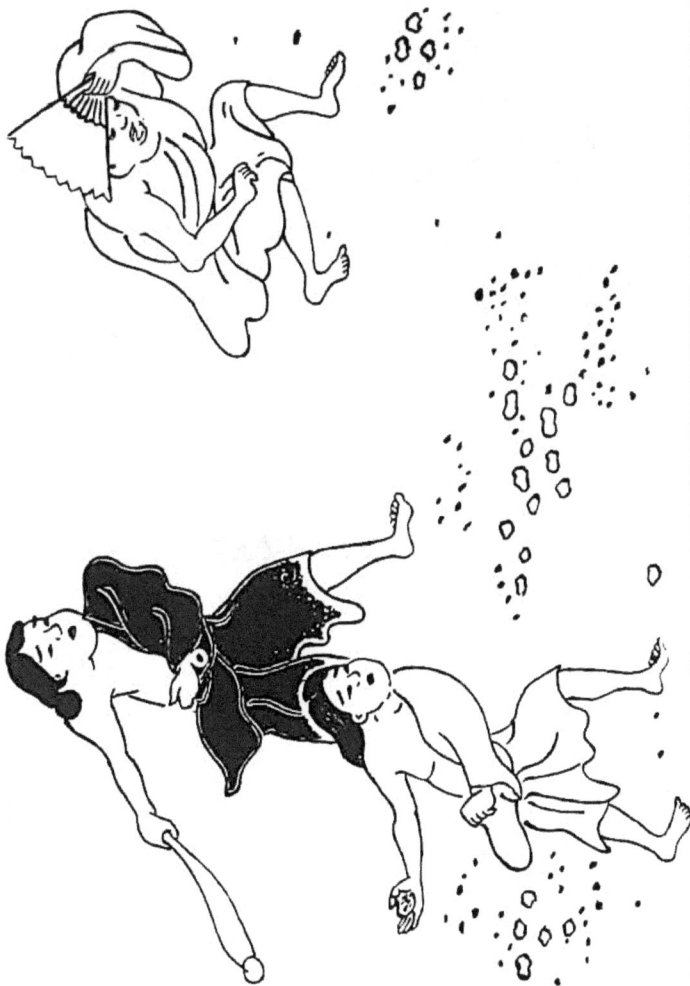

Illustration of a Rock War
Two teams or two villages would have a "playful" rock war that often turned quite serious. This undated illustration shows the rock war at Watarase River in Gunma Prefecture.

海國兵談所載

阿蘭陀人之國橖

Illustration of Dutch People Practicing Slinging Rocks at a Target Head
From Military Tales From Island Nations 1791

はしがき

武術には

素手で打ち突き蹴り組み投げ

絞め抑え合って闘う格闘武術と

武器となるべき得物を持って撃ち突き

斬り合って闘う撃突武術と

遠く離れた処から器具を使わず手で物

を投げつけて敵を倒す投擲武術と

弓鉄砲に類する機具を用いて射ち

撃って敵を斃す射撃武術とがある

本書はその武術のうち投擲武術特に

手裏剣術に関する一切の事を初伝より奥

伝印可まで図解説述したものである

藤田西湖

Introduction

One type of fighting is Sude, or the martial art of fighting barehanded. With this art you can hit, strike, kick, grapple, throw, choke or hold an opponent down. Another way to fight is to use Gekitotsu Jutsu or Attacking and Stabbing Martial Arts. This involves employing a specialized weapon to deliver a powerful impact, to stab or cut. This is called fighting directly with a weapon. In addition, there is a style of fighting which involves throwing something held in your hand to topple a distant enemy. This is done without any sort of implement to hurl the projectile. This method of attacking is called Toteki Bujutsu, the martial art of throwing.

While bows and rifles also hurl objects and kill opponents from a far distance, that art is called Shateki Bujutsu, or Shooting. This book will examine Toteki Bujutsu, with a particular focus on Shuriken Jutsu, the art of throwing Shuriken. I will present illustrated descriptions of these techniques from Shoden, the most fundamental level, and proceeding to Okuden, or expert level.

Fujita Seiko

投擲武術には、投擲術・投槍術・打根術・手裏剣術等がある。

投擲術（飛礫術・つぶての術・石抛術ともいう）は、敵に石を投げつけていためたおす術で、その方法には、素手で投げるのと、機具を用いて投げるのとがある。素手で石を投げる石投げ術は、投擲武術の最初のもので、後の投矛・投げ槍・打根・手裏剣術の元祖ともいうべきものであり、機具（石抛器）による石抛げ法は、射撃武術の元祖でもある。

投槍術とは、投げ槍用の短槍（普通の手槍を用いることもある）をもって、敵を突いたり、投げつけて倒す術で、この投げ槍術が、手突矢、投矢（打根）となったようである。この投げ槍の一種は、銛、籍といって、今でも狩漁猟に用いられている。

打根術（討根）、手突矢・投矢は、投げ突き用の短矢をもって、敵を突いたり、投げつけて敵を倒す術で、投げ槍術より進化したものである。

手裏剣術は、手裏剣を投げ打って敵を倒す術で、投擲・投槍・打根等の術より出で、さらに進化したものである。

10

What is Toteki Jutsu, the Art of Throwing?

The martial art of to 投擲術 Toteki Jutsu is comprised of the following:
● Tsubute Jutsu
● Toso Jutsu
● Uchine Jutsu
● Shuriken Jutsu

Tsubute Jutsu, or The Art of Throwing Stones, can be written in Kanji as follows:
● 飛礫術
● つぶての術
● 石抛術

    Tsubute Jutsu is the art of throwing rocks at your enemy. There are two ways this can be done, either barehanded or using a device to launch the rock. Sekinage Jutsu, or throwing rocks barehanded, is the earliest kind of Toseki Jutsu. Later throwing lances, spears, darts and Shuriken evolved from this. Similarly using catapults to hurl rocks in the earliest iteration of what later became Shateki Jutsu, the shooting arts.

    Toso Jutsu, the art of spear throwing, describes a method of throwing the short spear and impaling the enemy. The weapon used is a Teyari, or "hand spear," which is a short spear. Later, the Uchine, the Japanese throwing dart, evolved from Toso Jutsu.
One type of spear called a Mori, or harpoon, is still being used by those that hunt and fish.

    Uchine Jutsu is the art of using the Japanese throwing dart. These are thrown by hand to impale the enemy. Uchine can be written as 打根 "striking shaft" or 討根 "attacking shaft." Other words for the same weapon are Tezuki Ya "Hand stabbing Arrow" and Toya "Throwing Arrow."

    Shuriken Jutsu is the art of throwing Shuriken in order to topple your enemy. It evolved out of rock throwing, spear throwing and Japanese dart throwing. It then became more refined.

つぶて術

Tsubute Jutsu

The Art of Throwing Stones

石投術

　石を投げ当てるということは、やさしいようでなかなかむつかしいものである。遠近によって、目標のネライ方・投げ方があり、投げる石の大小・軽重・円丸・方角・扁平によっても、みなそれぞれ持ち方投げ方があり、一様でない。目標も、動かぬ物に投げ当てることは、さして至難のわざではないが、動く物・飛ぶもの・下がるもの・進み来るもの・走り去るものに、的確に投げ当てることはなかなかむつかしいものである。

13

Tsubute Jutsu, also known as Ishi Hajiki Jutsu
The Art of Throwing Stones

While it may seem easy to pick up a rock and throw it, Tsubute Jutsu is actually quite complex. The spot you aim for and the way you throw changes based on the distance. There is also no set way to throw a stone. The way you hold the rock when throwing depends on the size, weight, roundness, flatness, length and sharpness of the edges. It is not especially hard to throw a rock and hit a stationary object, however striking a target that is moving, flying, descending, advancing or fleeing is quite difficult.

目標のネライ方

目標よりすべて上方をネラッて、手は目標の中点まで下げるように投げる。たとえば、敵の顔面に当てようとする場合、手はアゴの下あたりまで下げるように投げるのが要領である。

投げ方も、手先だけで投げるようではいけない。肩・腕・手を目標にむかって一直線に、全身の力を一つにして、突き込むように投げるべきである。その一つでも狂えば、決して遠くえも届かず、目標にも当たらず、当たっても力なく、きくものではない。

走るものに対しては、その走る方向に、追うように横に投げるのであるが、遠近・速力の度合によって遅速があってはいけない。上にあがるものに対しては、また上から下がってくるものに対しても、その速力・度合によっておのずから手加減がある。

左方はよいが、右方え転ずるものは具合が悪いものである。身体の開き度合によって投げるべきである。

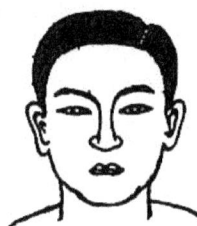

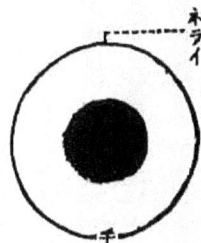

Mokuhyo no Neraikata : How to Aim

You should always aim to hit above your actual target while your hand should drop down below your target as you throw. For example, if you want to strike an enemy in the face, your hand should drop down to his chin as you throw.

Also, when you throw, only using the ends of your fingers is wrong. Your shoulder, arm and hand should all move in unison directly at your target. You should throw with full power as if you are trying to impale the enemy with the force of this action. If any of these aspects is out of sync then your throw will not travel far enough, will not strike your intended target and lack power therefore being ineffective.

If you are throwing at a target that is running, you need to throw across his path as if chasing him from behind. However, you have to be able to properly judge the distance to and speed of your target. This applies to targets above you or targets descending towards you. You have to be able to judge that target's speed and adjust accordingly. It is a fairly simple matter to hit targets on your left but targets on your right can be problematic. It requires you to switch your stance and then throw.

持ち方の要領

丸めの物

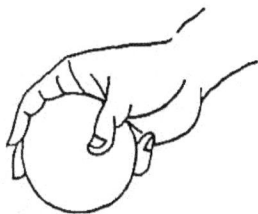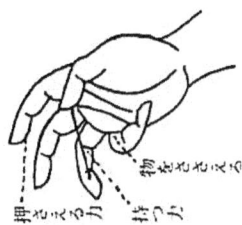

押さえる力　持つ力　物をささえる

やや大きな物

前と同じ

小さめの物

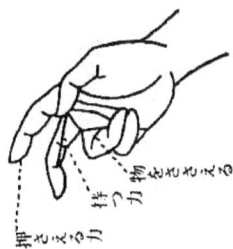

押さえる力　持つ力　物をささえる

長めの物

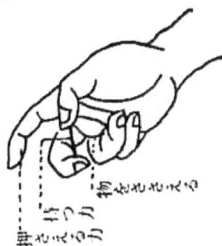

押さえる力　持つ力　物をささえる

17

How the fingers should grip different size stones for throwing.

Round rock

Pushing Power

Grip

Supporting Power

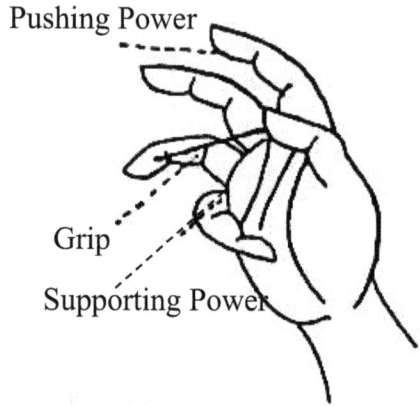

Slightly large rock

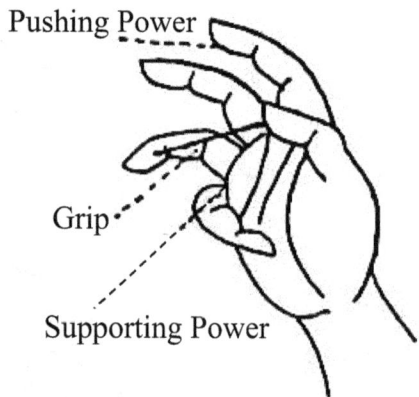

Pushing Power

Grip

Supporting Power

How the fingers should grip different size stones for throwing.

Small rock

Pushing Power

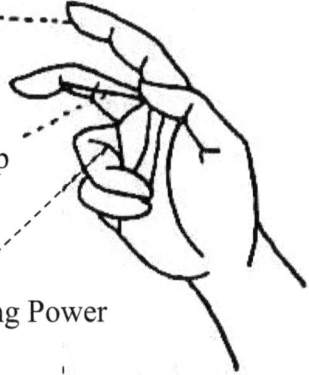

Grip

Supporting Power

Oblong rock

Pushing Power

Grip

Supporting
power

投げ方

物は投げた場所よりすぐ下にさがりがちのものである。

ネライの場所に物を投げ当てるには、投げる角度(体勢)で調子を取るか、手首で調子を取るべきものである。目標と体位が正確でないと外れるし、体位が正確であっても投げ方がもちろん外れる。

目付を正しく、体位を正方向でつくり、目標に物を突き込むような気持で投げつける。(この場合、物の持ち方と手離れに注意することが肝要)

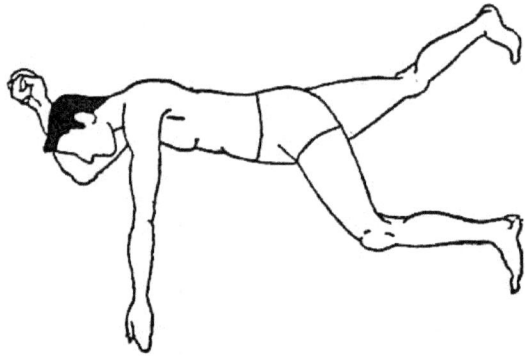

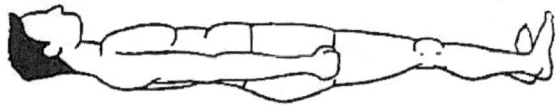

20

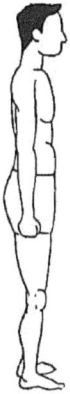

Toteki Jutsu
Nage Kata : How to Throw

When throwing things there is a tendency for your projectile to strike below the point you aim at. In order to strike your target accurately you need to combine the correct angle, as in body positioning, as well as releasing your projectile at the proper point. You will miss if your body positioning is off in relation to your target. Even having good body positioning can result in a miss if your throwing method is incorrect.

To throw properly you need to understand how to aim, stand with your foot facing your target and throw as if you are trying to impale it. In this situation it is important to focus on how you hold your projectile and the moment of release.

投げ方の要領

How to throw

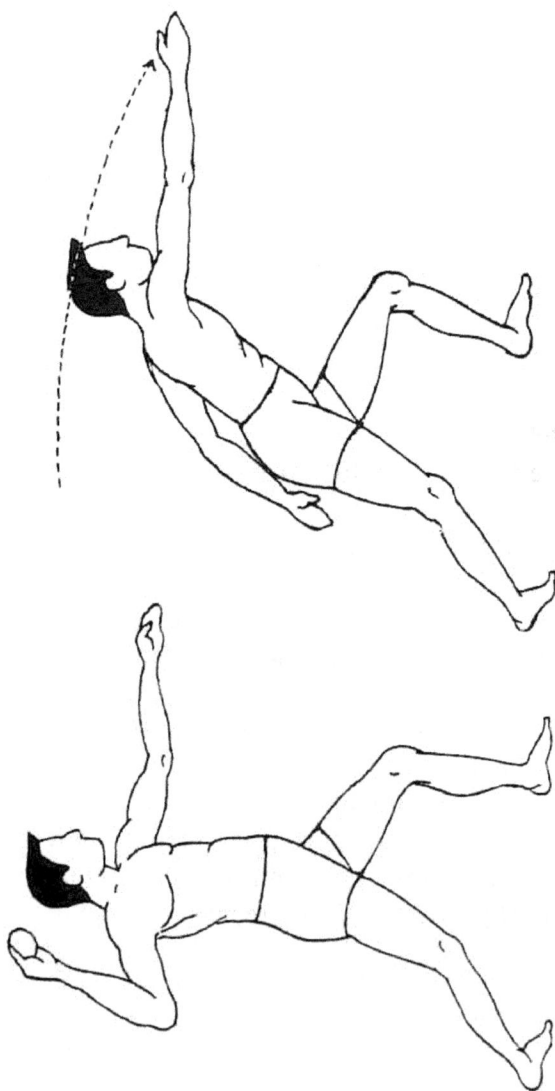

投槍

Toso

The Art of Throwing Spears

投 槍

武 具

鏢 鎗

投 矛

投 槍

漁 具

銛・鋸

簎

簎

投槍

武具

鏢鎗

投矛

投槍

猟具

銛り・矠り

簎

簎

Toso : Throwing Spears
Weapons

 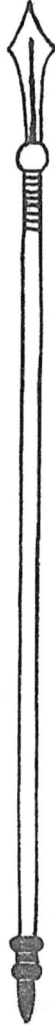 

投槍　投矛　鏢鎗　武具

Left: *Toso* Throwing Spear
Center : *Toma* Throwing Lance
Right: *Hyoso* Throwing Spear

Toso : Throwing Spears
Hunting Tools

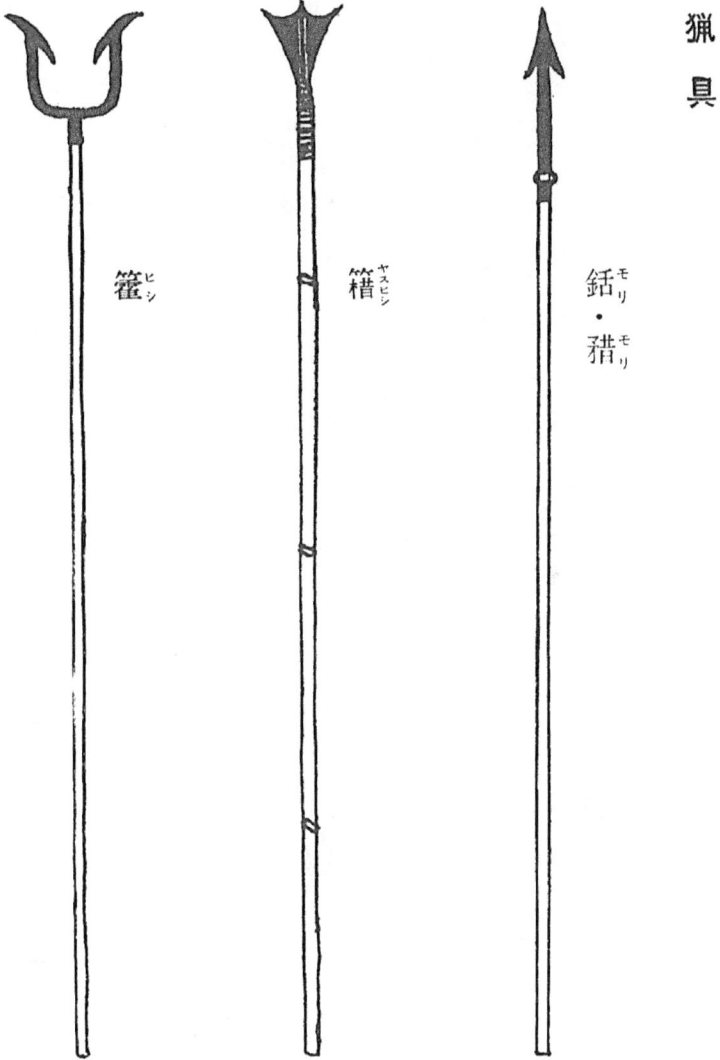

猟具

篊 ヒシ

簎 ヤスビシ

銛 モリ・猟 モリ

Left: *Bishi* Fishing Spear
Center : *Hakobishi* Fishing Lance
Right: *Mori* Fishing Spear

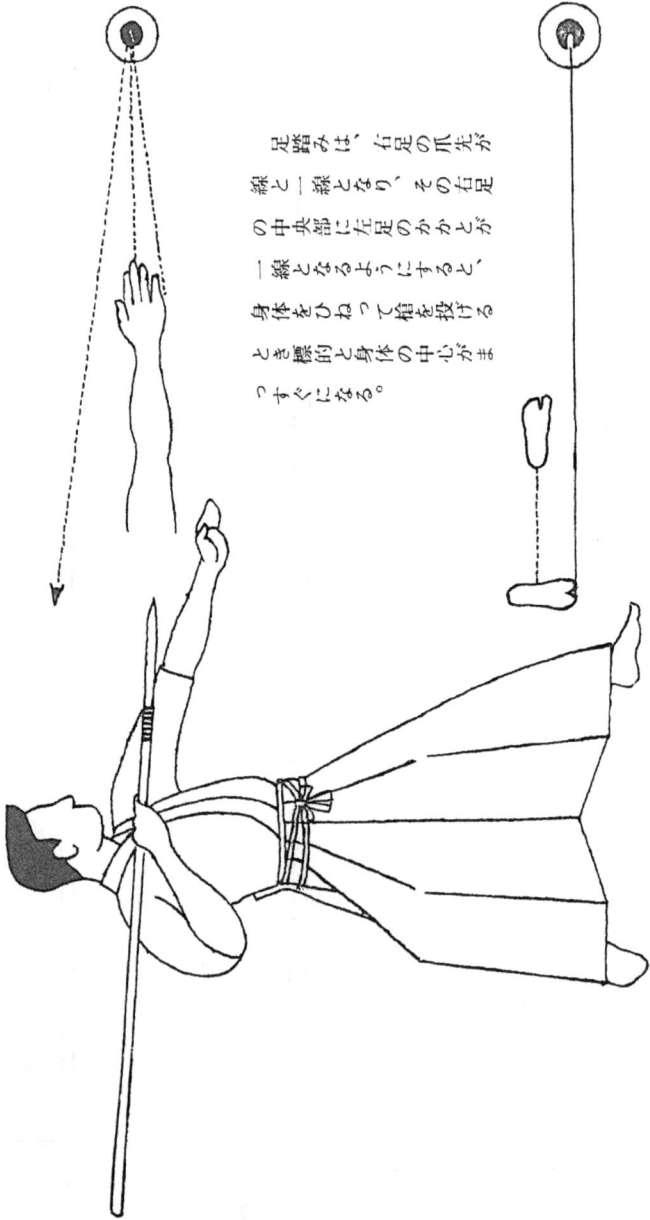

足踏みは、右足の爪先が
線と一線となり、その右足
の中央部に左足のかとが
一線となるようにすると、
身体をひねって槍を投げる
とき標的と身体の中心がま
っすぐになる。

足踏みは、右足の爪先が
線と一線となり、その右足
の中央部に左足のかかとが
一線となるようにすると、
身体をひねって槍を投げる
とき標的と身体の中心がま
っすぐになる。

How to throw a spear

If you draw a line from the tip of the toes of your right foot, it would end dead center of the target. Your left foot is parallel to this line, and if you were to draw a straight line from the back of the left heel it would intersect the center of your right foot. As you step forward with your right foot, your body twists and as you release your body is aligned with the center of the target.

打根

Uchine

Japanese Throwing Arrow

筈ハ竹ハマタハ堅木　太サ二寸五分クライ　長サ一尺クラ
イヨリ一尺五六寸クライマデ　羽ハ直羽三ツ立・四ツ立

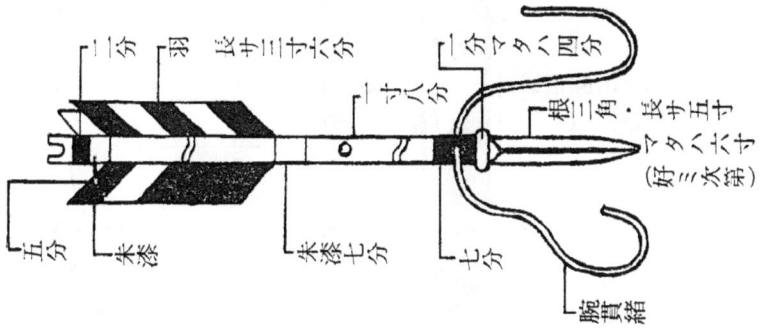

一分　羽　長サ三寸六分　一寸八分　一分マタハ四分

根三角・長サ五六寸
（好次第三）

五分　朱漆　朱漆七分　七分

腕貫緒

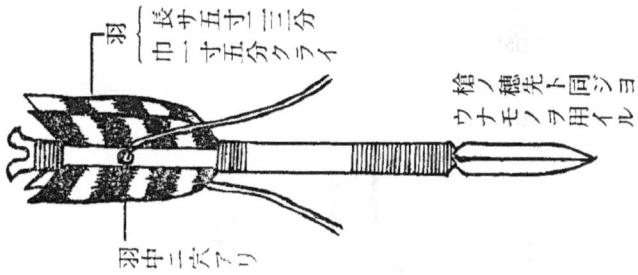

羽　長サ五寸三分　巾一寸五分クライ

穂先ト同ジヨ結ウナノトヲ用ル

羽中三六アリ

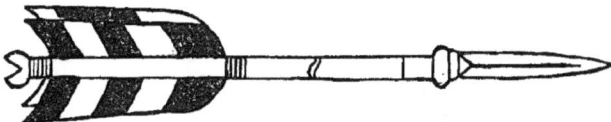

31

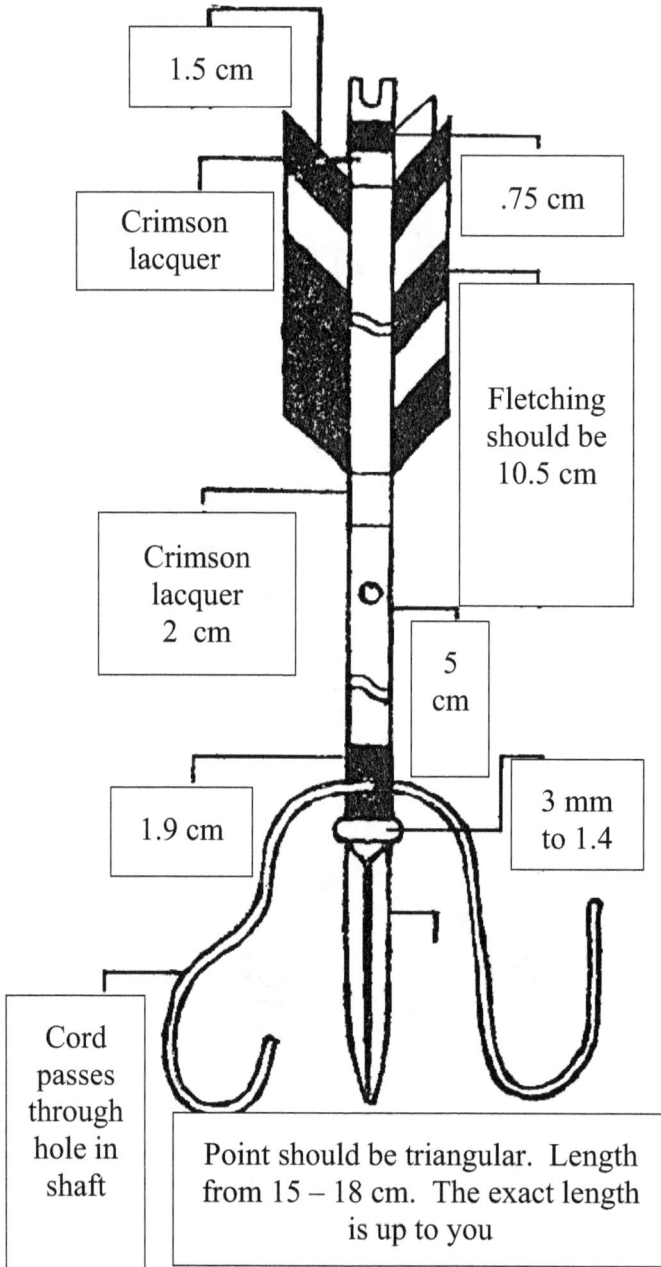

1.5 cm

Crimson lacquer

.75 cm

Crimson lacquer 2 cm

Fletching should be 10.5 cm

5 cm

1.9 cm

3 mm to 1.4

Cord passes through hole in shaft

Point should be triangular. Length from 15 – 18 cm. The exact length is up to you

筬ハ竹マタハ堅木、太サ二寸五分クライ、長サ一尺クラ

イヨリ一尺五六寸クライマデ、羽ハ直羽三ツ立・四ツ立

The shaft can be either bamboo or oak. The thickness is around 7.5 cm and the length from around 30 ~ 45 cm. The feathers are cut straight and either 3 or 4 are used.

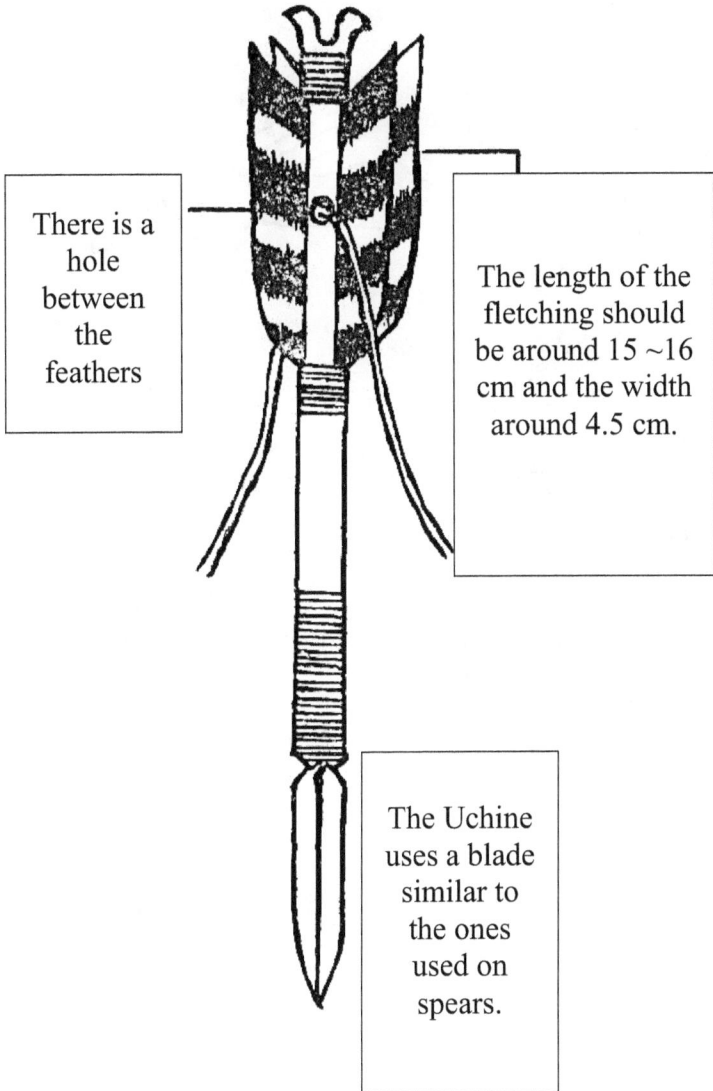

There is a hole between the feathers

The length of the fletching should be around 15 ~16 cm and the width around 4.5 cm.

The Uchine uses a blade similar to the ones used on spears.

打根の持ち方

構え

突き

投げ

Ways to hold and strike with the Uchine:

Left: *Nage* - Throw   Center: *Tsuki* - Stabbing   Right: *Kamae* - Stance

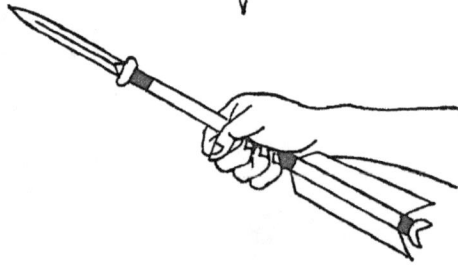

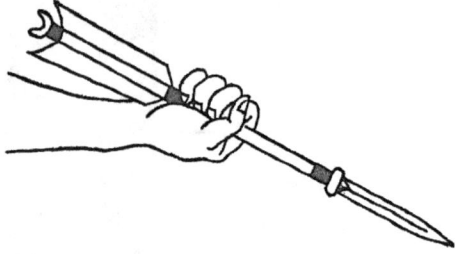

打根用法
十五箇条

Using the Uchine,
Japanese Throwing Arrow

•

Fifteen Techniques

- *Tachi-ai.* Three techniques for use against an opponent armed with a Tachi.

- *Yari-ai • Naginata-ai.* Three Techniques for use against opponents armed with a spear or halberd.

- *Yumizuke.* "Fixing to a Bow." Five techniques to use when your bowstring breaks, using just the arrow.

- *Anya.* Four techniques for fighting at night or in the dark.

打根用法十五箇条

投擲武術

太刀合三箇条

一　附入

打根を右構えに持ち
敵の打ち出す太刀を
払ひ附け入るなり

二二

● *Tachi-ai.* **Three techniques for use against an opponent armed with a Tachi.**

#1
Tsuke Iri
Enter Close

Hold the Uchine in a Migi Kamae, or in your right hand with your right side facing forward. When the opponent cuts down at you, sweep away in movement called Harai and then stab him.

二　受流

打根を下げ打ち込む
太刀を受け流して附
け入るなり

#2
Uke Nagashi
Receive and Pass

Hold the Uchine facing downward. The opponent cuts. Receive the downward swinging sword and allow it to pass by you, then stab the opponent.

三　打込

手近なる敵故打根を
右の肩に担ぎ打ち附
くるなり

#3
Uchi Komi
Move in and Strike

When an enemy is close by, bring the Uchine up to your shoulder and then strike down.

槍　長刀合三箇条

なり
越えて敵の体を突く
突出す槍長刀を飛び
四　飛乱

- **Yari-ai · Naginata-ai. Three Techniques for use against opponents armed with a spear or halberd.**

#4
Hiran
Wild Leap

When the opponent thrusts at you with a spear or halberd, leap forward and stab the opponent's body.

五　留手

打根を青眼に構え突
出す槍長刀を払ひ附
くるなり

#5
Tomete
Stopping Hand

Hold the Uchine in Seigan no Kamae, with the right side of your body facing the opponent and all your concentration on the opponent's eyes. When he stabs with the spear or halberd, sweep it aside and stab.

六　水月

打根を左手に持ち突
き出し薙ぎ払ふ長刀
を受け流しつゝ附け
入るなり

#6
Suigetsu
The Moon Reflected on Water

Hold the Uchine in your left hand with the point facing the enemy.
When your opponent launches a sweeping cut with his halberd,
receive it and allow it to pass by, then stab.

弓附（弦切、矢尽）五箇条

七　打落

打根を右手に持ち槍
刀を突き出し斬り附
くるを打ち払ひ落し
附くるなり

- **_Yumizuke._ "Fixing to a Bow."** Five techniques to use when your bowstring breaks, using just the arrow.

#7
Uchi Otoshi
Striking and Dropping

Hold the Uchine in your right hand. When the opponent stabs or swings with his halberd or katana strike down in a hard-sweeping motion causing the opponent to drop his weapon.

八　突上

打根を右手に持ち青
眼に構え附け込みて
敵の脇腹を突くなり

#8
Tsuki Age
Upward Stab

Hold the Uchine in Seigan no Kamae, with the right side of your
body facing the opponent. Focus your energy on the opponent's
eye. As soon as you see an opening, dart in and stab to the
opponent's abdomen on the side.

九　受留

切り込み附け込む頭
を打根にて払ひ又は
留めて附け入るなり

#9
Uketome
Receive and stop

Wait until the opponent tries to cut your head. Use the Uchine to sweep his sword away or block it and then stab.

十　払留

太刀槍其外とも敵の
得物を打ち留め払ふ
ことなり

#10
Harai Tome
Sweep and Stop

No matter what weapon the opponent is armed with - spear, halberd
or other weapon, sweep it away or stop it then stab.

十一　柳露

突き出す槍を飛び越
えて打根を投げ附く
るなり

#11
Ryuro
Mist on a Willow Branch

Leap away and throw the Uchine at the opponent the moment he
stabs with his spear.

暗夜四箇条

十二　透目附

能く透し見て静かに
歩み寄り敵のすきを
突くなり

● *Anya*. **Four techniques for fighting at night or in the dark.**

#12
Sukine Tsuke
Attacking the Opening

Maximize your awareness of your surroundings and step carefully toward the enemy. Finding an opening, stab.

打根を左手に持ち下
して能く心を沈め之
れも静かに進みて敵
の容子を見て附け入
るなり

十三　静心

#13
Seishin
Silent Spirit

Hold the Uchine in your left hand with the point facing downward.
You should completely calm yourself and eliminate any tension in
your body. Find a gap in the opponent's guard and then stab.

十四　探突

文字の如く左右遠近
をうかゞひ附け入り
て敵の油断を見て刺
すなり

#14
Tantotsu
Seeking a Place to Stab

Just as the meaning of the Kanji used imply, shift your body left and right, closer or farther from the enemy. Your goal is causing the opponent's guard to slip. When the opponent leaves himself open, stab.

十五　捨見

右手に打根を下げて
敵の居所を見定め早
足に進みて突くなり

#15
Shashin
Sacrifice Yourself
Note: This Kanji combination is more commonly read as Sutemi.

Holding the Uchine in your right hand carefully establish the distance between you and the opponent. Then dash in as rapidly as possible and stab.

打根の打ち方

要領は、投縄と同じ。

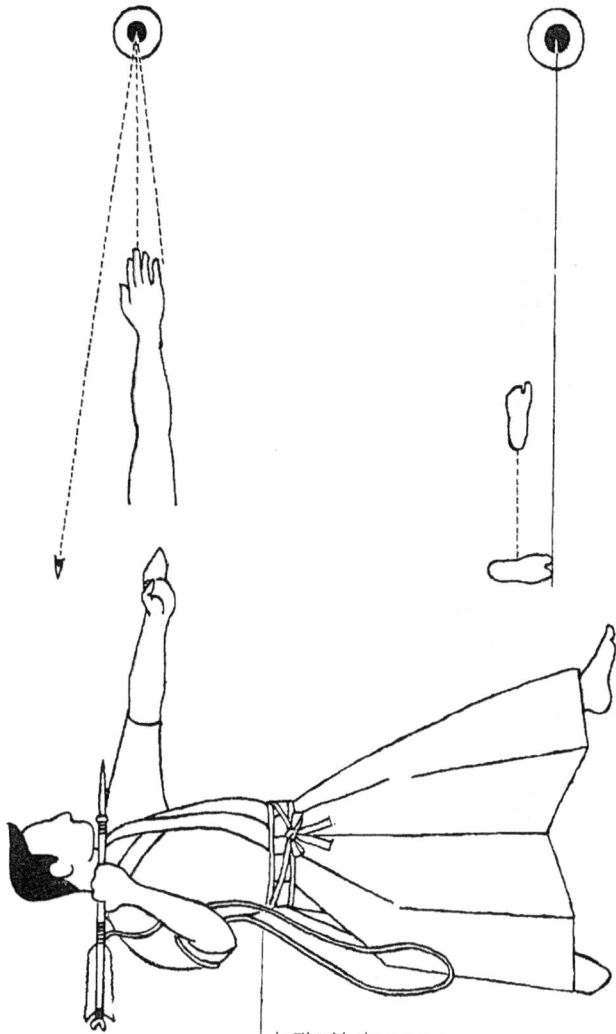

組んで打ちつける
打根を敵に打つとき
は手元に引きもどって

51

打根の打ち方
要領は、投槍と同じ。

打根を敵に打ちつけては下元に引きもどす紐

Uchine no Uchi-kata
How to Throw The Uchine

Your hips should be positioned in the same way as when you throw a javelin. The cord attached to your elbow is used to retrieve the Uchine after it has struck the enemy.

支那袖箭その他

支那には'袖箭'流星箭'鞭箭'筒子箭がある。

袖箭は'短かく袖にかくしていて手で投げる'三十歩くらいの敵を倒すことができる。わが国ではこれを'打根または手裏矢という。

袖箭

流星箭も同法は同じであるが'袖箭よりも先が重い。

流星箭

鞭箭は'鋼を溜子にして放す。

鞭箭

筒子箭は'竹の筒の中え長さ一尺二寸くらいの短箭を入れて'手で発するもので'箭に毒薬のついたものもある。わが国の吹矢と手裏剣を合わせたもの。

筒子箭

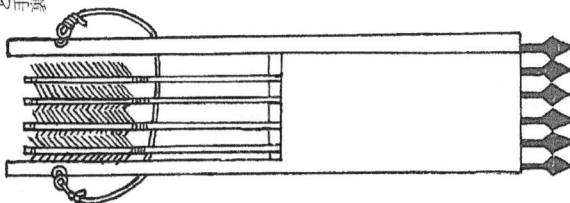

China's Sleeve Dart and Other Related Weapons

China developed weapons similar to the Uchine such as the Sleeve Dart, the Shooting Star Arrow, the Whip Arrow and the Tube Arrow.

- The Sleeve Dart is short and can be hid up the sleeve. It is accurate up to thirty paces. This is called Hand Spear or Uchine in our country.

- The Shooting Star Arrow is used in the same fashion as the Sleeve Dart though the arrowhead is heavier.

- The Whip Arrow is flung from a launcher.

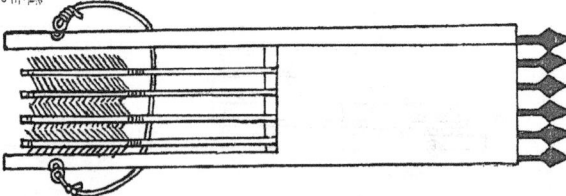

- The Tube Arrows are fitted inside a bamboo case. The arrows are roughly 36 cm in length and are launched by hand. The tips are sometimes coated with poison. With reference to Japanese weapons, this is sort of a combination of blowdarts and Shuriken.

手裏剣

Shuriken

日 本 武 尊

Prince Yamato Takeru (71 – 114)

手裏剣術とは

手裏剣術のはじめは、人皇第十二代景行天皇（六四八）の皇子日本武尊（七七三）が御東征の帰途、足柄（今の神奈川県—古事記）坂下において（日本書紀には信濃の国となっている）糧食をとられておられたき、その坂の神が、白い鹿になって来かかった。そのとき尊は、食べ残しの蒜（ヒル）の片端をもって、その白鹿の目に打ち当て、打ち殺されたという古事記、日本書紀の事跡をもって始めとされる。

手裏剣術とは、遠く離れたところから、手の裡（裏）の剣（陰剣）を敵に、投げ、撃ち、離して倒し、勝理（利）を修むるわざ（術）という意義で名づけられたもので、したがって名称も、手裡剣術・手裏剣術・手離剣術・投剣術・打剣術・撃剣術・修理剣・修利剣術とも書かれているわけである。

支那では、銑鋧・鏢・三不過術といっている。

この手裏剣術は、昔まだ飛道具の発達しなかった頃は、武士は飛道具にかわ

What are Shuriken?

Shuriken originated with Prince Yamato Takeru (71-114 AD) the son of Emperor Keiko, the 12th Emperor of Japan (71-130 AD.) One day, Prince Takeru was returning from subjugating the barbarians in the east when he stopped at Ashigara Pass. The Kojiki, or *Records of Ancient Matters,* written in 711-712 AD, states this was near present day Kanagawa Prefecture. The Nihonshoki, or *The Chronicles of Japan*, published in 720 AD, states that this was in Mino Domain. As he was gathering food the Deity of the Slope transformed into a white deer and approached him. Scooping up some of his leftover meal in his hand, he threw scraps of Nira garlic chives into the eyes of the deer, killing it. This incident, which is the first use of Shuriken, is recorded in both the *Records of Ancient Matter* and *Chronicles of Japan.*

Shuriken Jutsu is the art of throwing a blade concealed in the palm at a distant enemy. The Kanji are Hand + Reverse + Sword. So a sword hidden in the reverse, or palm, of the hand. Other Kanji combinations are possible including ones that mean "a way to achieve victory." Other Kanji combinations are:

手裡劍術 Hand + Behind + Sword + Technique
手裏劍術 Hand + Reverse + Sword + Technique
手離劍術 Hand + Release + Sword + Technique
投劍術　 Throw + Sword + Technique
打劍術　 Strike + Sword + Technique
擊劍術　 Violent + Sword + Technique
修理劍　 Achieve Victory + Sword
修利劍術 Achieve Victory + Sword + Technique

In China Shuriken are known as 鉄鋧 Shuken or 鏢 Hyo and the art of Shuriken can be called 三不過術 Sanbuka Jutsu

The Shuriken arts began, in their most primitive form, long ago. Thrown weapons were one of the techniques studied by Samurai, in conjunction with other martial arts.

るものの一つとして、他の武術と共に学んだものである。

この手裏剣術は、護身と攻撃を兼ねた術で、大別すると二法となる。

その一を留手裏剣、他を貴手裏剣という。

留手裏剣には、

忍手裏剣・静定剣・乱定剣

の三伝があり

貴手裏剣には

火勢剣・薬剣・毒剣

の三伝がある。

忍手裏剣というのは、手裏剣術用として特に用意された（特定の）手裏剣をもって、敵を撃つ方法で、通常いわゆる手裏剣術とは、これをいうのである。

その武器として用いる手裏剣の形態は種々あり、一様でない。流派々々によっても、一種独特の手裏剣形態がある。

針形・棒状・角形・釘形・平形・短刀形・槍の穂形等のほか、投げつければどこか一角か二角は必らず突き刺さるように造った三光・四方・星状（五方）・六方・八方・十方・十字・車剣・万字形等がおる。これらは、総称してすべて車剣という法輪より出たものである。

Shuriken Jutsu was considered to be both a self-defense and an offensive art. Shuriken Jutsu is broadly divided into two categories:

Tome Shuriken Jutsu : Stopping Shuriken
Seme Shuriken Jutsu : Punishment Shuriken

Stopping Shuriken consists of the following three :
忍手裏剣 Shinobi Shuriken : Ninja Shuriken
静定剣 Seiteiken : Static Object Shuriken
乱定剣 Ranteiken : Violent Shuriken

Punishment Shuriken consists of the following three :
火勢剣 Hiseiken : Fire Effect Blades
薬剣 Yakuken : Medicinal Blades
毒剣 Dokuken : Poisoned Blades

Shinobi Shuriken describes a weapon specifically crafted to be used when attacking the enemy as well as the techniques employed. This is what is generally referred to as Shuriken Jutsu. The size and shape of the Shuriken being employed can vary widely and there is no set shape. Each Ryuha, or school of martial arts, will have its own particular style of Shuriken.

Some of those shapes are :

針形 Hari Gata : Needle Shaped
棒状 Bo-Joh : Stick Shaped
角形 Kaku Gata : Faceted
釘形 Kugi Gata : Nail Shaped
平形 Hira Gata : Flat Shaped
短刀形 Tanto Gata : Knife Shaped
槍の穂形 Yari no Ho Gata : Spear-point Shaped

手裏剣打法（棒状の）には、直打法と廻転法の二種があり、剣先を指先の方にして打つのを直打法、剣先を掌の中にして打つのを廻転法という。

打ち方には、正常打・横打・逆打の三法がある。投撃を加える個処は、眉間・両眼・人中・咽喉・心臓部・乳部・水落・脇腹・臍部としている。

静定剣とは、特定の手裏剣を用いず、とっさの場合に、特ちあわす小刀・小柄・笄等をもって手裏剣代用として撃つ方法である。よく「小柄を抜いて手裏剣として撃つ」等といわれるのは、この静定剣のことである。したがって武士は心得としても、常に手裏剣術を練磨したものである。

乱定剣とは、急場にのぞみ、有合う器物、たとえば、火入れ・灰吹き・盆・茶碗・鉄瓶等何んでもそこに有合う物品を敵に投げ付けて、急場の危地を脱する方法である。

In addition, there are Shuriken crafted so that two or three points will always stick in the target. Some of these are :

三光 San Koh : Three Prongs of Light
四方 Shiho : Four Direction
星状 Hoshijo : Five-Pointed Star
六形 Roku Gata : Six Pointed
八形 Hachi Gata : Eight Pointed
十形 Ju Gata : Ten Pointed
十字 Juji : Cross Shaped
車剣 Kuruma Ken :
万字型 Manji Kata : 10,000 Shaped

Translator's Note:
"10,000 Shaped" is another way to refer to a Swastika shape 卍. The Manji, or Swastika, is meant to be seen as a wheel, rotating freely in all three dimensions.

These are all generally referred to as Kuruma Ken, or "wheel swords" which are based on the shape of the Dharma Chakra "Wheel of the Law." There are three ways to throw a Bo Kata, or stick-shaped, Shuriken :

正常打 Seijo Uchi : Standard Throw
横打 Yoko Uchi : Side Throw
逆打 Gyaku Uchi : Reverse Throw

The targets you aim for are as follows :

眉間 Miken : Between the Eyebrows
両眼 Ryogn : Both Eyes
人中 Jinchu : A Kyusho, striking point, located below the nose
心臓部 Shinzobu : Near the heart
乳部 Nyubu : Area around the nipples
水落 Mizuochi : The Solar Plexus
脇腹 Wakibara : The Armpit
臍部 Saibu: Navel

Seiteiken Static Object Shuriken

Seiteiken, Static Object Shuriken techniques, does not describe a specific type of Shuriken, instead this is the art of using any static object on your person as a Shuriken. This includes throwing short swords, knives, Kozuka utility knives and so on when in an extreme situation. You often hear the phrase, "I drew my Kozuka utility knife and threw it like a Shuriken." This is what is meant by Silent Shuriken. In other words, throwing Shuriken was something every Samurai was prepared to do, and they trained in this art extensively.

Ranteiken Violent Shuriken

If you are in an extreme situation then you would employ Ranteiken, Violent Shuriken, which is using anything at hand as a Shuriken. For example a fire pan full of hot coals, a bamboo ash holder, a pot, a tea bowl an iron kettle or any other such object. Throwing any such object in the enemy's face will allow you to escape from a dangerous situation. This is Ranteiken.

貴手裏剣に属する火勢剣とは、火矢・松明・ほうろく火等をもって、敵を責める方法で、今日の擲弾筒（手榴弾）を投げる等は、この火勢剣である。

薬剣（不殺必倒の剣ともいう）は、目潰し（遠当術）・霞扇の術・または息討器による方法等である。この方法は、敵を殺さず、仮死せしむるにある。多くは大事な敵を捕縛するとき等に用いた方法で、器具・薬法等十数通りある。

毒剣（必殺不生の剣ともいう）は、どうしても倒さねばならぬ敵に用いる方法で、多くは手に負えぬような剛敵を倒すのに用いた方法である。瞬時に即死せしむるのと、数時間後に死に至らしむるのとによって、その器具・薬法にそれぞれの秘伝がある。

Seme Shuriken **Attacking Shuriken**

Seme Shuriken, or Attacking Shuriken, contains Hiseiken, Flaming Blades. This includes attacking the enemy with fire arrows, torches or Horokuhi, gunpowder launched fire arrows. These Hiseiken are similar to today's grenade launchers or hand grenades.

Yakuken **Medicine Blades**

Yakuken, or Medicine Blades, are also referred to as "Non-lethal blades that will always fell the enemy." These consist of Blinding powders, which are part of "Attacking at a Distance Techniques," "Fanning the Mist Techniques" and "Breath Stoppers." Each of these describes a separate way to attack the enemy without killing them. Instead, it renders them unconscious. Many of these techniques are used when capturing and tying up an important enemy official. There are around ten different implements and methods used in Medicine Blades.

Dokuken **Poisoned Blades**

Dokuken, Poisoned Blades that will always fell the enemy, are used when faced with an adversary that you must eliminate. This method is typically used to topple a powerful opponent without being injured yourself. There are many different secret methods in Kokuken. Some of them kill instantly while others can take several hours. It all depends on the weapon used and how the medicine is prepared.

手裏剣流派

Encyclopedia

of

Shuriken

Schools

Translator's Note: The schools are listed according to the Japanese alphabet. Many schools of martial arts have Shuriken techniques in addition to other weapons however in the following list the schools that are primarily Shuriken are underlined. Some of the schools contain the year it was founded, while others do not.

あ

浅山一伝流　慶長　　浅山一伝斎重晨

○天津流　江戸初期　　天津小源太

荒木流　天正　　荒木夢仁斎源秀綱

い

伊賀流　永禄

○伊豆流　明和安永頃　　上遠野伊豆広秀

一刀流　　渋木新十郎

一方流　寛永　　難波一方斎藤原久長

え

円明流　慶長　　宮本武蔵政名

お

温古知新流　　川澄平九郎政光

か

春日流

○上遠野流　明和安永頃　　上遠野伊豆広秀

香取神刀流　　飯篠長威斎家直

鹿立流　寛永十七年云フ　　松林左馬助永吉

く

日下流　　日下一甫

こ

甲賀流　永禄

66

Schools that have Shuriken as their primary art are <u>underlined</u>.

Starting with あ A (Pronounced "Ah")

- **Asayama Ichiden School** 浅山一伝流 Founded by Asayama Ichidensai Shigetatsu 浅山一伝斎重晨 in the Keicho Era (1596 –1615.)
- <u>**Amatsu School** 天津流</u> Founded by Amatsu Kogenda 天津小源太 in the Early Edo Era (1600 – 1700.)
- **Araki School** 荒木流 Founded Araki Mujinsai Minamotono Hidetsuna 荒木夢仁斎源秀縄 in the Tensho Era (1573 – 1592.)

Starting with い I (Pronounced "E")

- **Iga School** 伊賀流 Founded in the Eiroku Era (1558 – 1570.)
- <u>**Izu School** 伊豆流</u> Founded by Kadono Hirohide 上遠野広秀 sometime in the Meiwa and An'ei Eras (1764 – 1781.)
- **Ikkan School** 一貫流 Founded by Shibuki Shinjuro 渋木新十郎
- **Ichikata School** 一方流 Founded by Nanaba Ichikata Saito Genhisagana 難波一方斎藤原久長 in the Kanei Era (1624 – 1644.)

Starting with え E (pronounced "Eh")

- **Enmei School** 円明流 Founded by Miyamoto Musashi Masana 宮本武蔵政名 in the Keicho Era (1596 –1615.)

Starting with お O

- **Onkochishin School** 温故知新流 Founded by Kawasumi Hirakyuro Masahiko 川澄平九郎政光

- **Kasukabe School** 春日部流
- **Kadono School** 上遠野流 Founded by Kadono Hirohide 上遠野広秀 sometime in the Meiwa and An'ei Eras (1764 – 1781.)
- **Katorishinto School** 香取神刀流 Founded by Izasa Choisai 飯篠長威斎家直
- **Ganryu School** 願立流 Also written without the second Kanji. Founded by Matsubayashi Samanosuke Nagayoshi 松林左馬之助永吉 Kanei Era (1624 – 1644.)

- **Kusaka School** 日下流 Founded by Kusaka Ippo 日下一甫.

- **Koga School** 甲賀流 Founded in the Eiroku Era (1558 – 1570.)
- **Koden "Teachings of the Fox" School** 狐伝流 Founded by Fujiwara no Kamatari 藤原鎌足
- **Kobori School** 小堀流 Founded by Kobori Kankai Yorinyudo Kiyohira Yoshiyuki 小堀勘解由入道清平好之

孤伝流　　　　　藤原鎌足

小堀流　　　　　小堀助解由人道湖清平好之

し

止心流　　　　　真杉三郎左衛門三設

実用流　文化　　平山行蔵潜

自得流　　　　　岩佐弥五左衛門済純

諸賀流　安永　　土川仁和右衛門至親

正雪流　　　　　由井民部助橘正雪

白井流　　　　　白井亨義謙

真陰流　天正　　疋田豊五郎景兼

○心月流　享保　藤原成忠

新心流　寛永　　関口八郎右衛門氏心

神道精武流　文化　小笠原城之助長政

せ

清心流　慶長　　森辺之助勝正

関口流　寛永　　関口八郎右衛門氏心

た

大東流　　　　　武田惣角

竹内一心流　　　篠原重右衛門一心斎藤原慶英

竹村流　　　　　竹村与右衛門玄利

立身流　　　　　立身三京

ち

知新流　正保　　飯篠市兵衛

69

- **Shishin School** 止心流 Founded by Masugi Saburo Saemon Misetsu 真杉三郎左衛門三設

- **Jitsuyo School** 実用流 Founded by Hirayama Kozo Heigen Sen 平山行藏潜 in the Bunka Era (1804 – 1818.)

- **Jitoku School** 自得流 Founded by Iwasa Yogo Saemon Kiyojun 岩佐弥五左衛門清純

- **Shosho School** 諸賞流 Founded by Tsuchikawa Ninnaemon Shishin 土川仁和衛門至親

- **Masayuki School** 正雪流 Founded by Yuikaki benosuke Tachibana no Masayuki 由井民部之助橘正雪

- **Shirai School** 白井流 Founded by Shirai Ko Giken 白井亨義謙

- **Shinkage School** 真陰流 Founded by Hikita Bungoro Kagetada 疋田豊五郎影兼 Tenshō Era (1573 – 1592)

- <u>**Shingetsu School** 心月流</u> Founded by Fujiwara Narachu 藤原成忠 in the Kyōhō Era (1716 – 1736)

- **Shinshin School** 神心流 Founded by Sekiguchi Hachiro Uemon Ujimune 関口弥六右衛門氏心 in the Kan'ei Era (1624 – 1644)

- **Shindo Seibu School** 神道精武流 Founded by Ogasawara Jonosuke Nagamasa 小笠原城助長政 in the Bunka Era (1804 – 1818)

- **Seishin School** 清心流 Founded by Mori Kasuminosuke Katsushige 森霞之助勝重 in the Keicho Era (1596 – 1615)

- **Sekiguchi School** 関口流 Founded by Sekiguchi Hachiro Uemon Ujimune 関口弥六右衛門氏心 in the Kan'ei Era (1624 – 1644)

Starting with た Ta

- **Daito School** 大東流 Founded by Takeda Sokaku 武田惣角
- **Takeuchi Isshin School** 竹内一心流 Founded by Shinohara Shigeuemon Isshinsai Fujiwara Yoshihide 篠原重右衛門一心斎藤原慶英
- **Takemura School** 竹村流 Founded by Takemura Youemon Genri 竹村与右衛門玄利
- **Tatsumi School** 立身流 Founded by Tatsumi Sankyo 立身三京

Starting with ち Chi

- **Chishin School** 知新流 Founded by Ishima Ichihyoei 飯島市兵衛 in the Shōhō Era (1644 – 1648)

つ

○津川流

て

天刀流　吸江十右衛門

天真伝一刀流　白井流ナリ　白井亨

に

○丹羽流　宝暦頃　丹羽織江氏張

わ

○根岸流　根岸忠誠宣教松齢

根来流

は

宝山流　元亀　堤山城守宝山

平集流

ま

松末流流　中野伴水景達

松巣流

み

三浦流　三浦揚心

- **Tsukawa School** 津川流

- **Ten "Heavenly" School** 天流 Founded by Gyuko Juemon 吸江十右衛門
- **Tenshinden Itto School** 天真伝一刀流 This school is also known as Shirai School 白井流 Founded by Shirai Ryo 白井享

- **Niwashi School** 丹羽流 Founded by Tanba Orie Ujiharu 丹羽織江氏張 in the Hōreki Era (1751 - 1764)

- **Negishi School** 根岸流 Founded by Negishi Chuzo Senkyo Shorei 根岸忠蔵宣教松齢
- **Nerai School** 根来流

- **Hozan School** 宝山流 Founded by Tsutsumi Yamashiro no Kami Hozan 堤山城守宝山 in the Genki Era (1570 – 1573)

Starting with へ He (Pronounced "heh")

- **Heishu School** 平集流

Starting with ま Ma

- **Matsu School** 末流 Founded by Nakano Tomomizu Kagetatsu 中野伴水景達
- **Matsuba School** 松葉流

Starting with み Mi

- **Miura School** 三浦流 Founded by Miura Yoshin 三浦揚心

む

武藏流　慶長　宮本武藏政名

も

○な洲流　　小紫惣兵衛

○毛利流　　毛利游太郎玄達

や

柳生流　慶長　柳生但馬守宗厳

山内流　　山内須藤用部秀久武休切

ゆ

融和流　　伊藤伴右衛門高豊

よ

○掃心流　慶長　秋山四郎左衛門義時

○義尾流

○印は手裏術を主とした流名

Starting with も Mo

- **Moen School** 孟淵流 Founded by Komurasaki Sohyoei 小紫惣兵衛
- **Mori School** 毛利流 Founded by Mori Gentaro Gentatsu 毛利源太郎玄達

Starting with や Ya

- **Yagyu School** 柳生流 Founded by Yagyu Tanba no Kami Muneyoshi 柳生但馬守宗厳
- **Yamauchi School** 山内流 Founded by Yamauchi Suto Keibu Hidehisa Mukyusai 山内須藤刑部秀久無休斎

Starting with ゆ Yu

- **Yuwa School** 融和流 Founded by Itoh Tomouemon Takatoyo 伊藤伴右衛門高豊

Starting with よ Yo

- **Yoshin School** 揚心流 Founded by Akiyama Shiro Saemon Yoshitoki 秋山四郎左衛門義時 in the Keicho Era (1596 – 1615)
- **Yoshio School** 義尾流

手裏剣術

伝書

Shuriken
Densho
Documents of
Transmission

Translator's note: Densho are instruction documents handed down within a martial arts tradition. The documents Fujita Seiko included in this book contain varying degrees of detail. Some Densho only list the names of the techniques and a list of previous inheritors of the school, while others may contain illustrations. Densho can also serve as certification of ability and proof the holder is a member of a certain school of Shuriken.

心月流手裏剣術目録

一、手裏剣軽重之事

一、手裏剣持様之事

一、手裏剣手之内之事

一、足踏之事

一、打出目付之事

一、指屈伸之事

一、手裏剣離之事

　右六箇条立打也

一、居打之事

一、左右打之事

一、前後打之事

一、歩行打之事

一、脇差懐剣打之事

一、闇夜之打様心得之事

一、手裏剣打様秘伝

　右三箇条有り

右之条々令相伝畢　於鍛錬修行有之者免許之口伝打方可令相伝者也仍目録之
慶長如件

藤原成忠

藤原義時

三三

Shingetsu Ryu Shuriken Jutsu Mokuroku
Catalogue of Shuriken Techniques For the Shingetsu School

- On the proper weight of a Shuriken
- On the proper way to hold a Shuriken
- The proper way to throw a Shuriken
- The proper way to stand
- The proper place to aim when throwing
- The proper way to apply pressure with the fingers
- The proper way to release the Shuriken

The above six categories are for throwing while standing

- On the subject of throwing while seated
- On the subject of throwing left and right hands
- On the subject of throwing ahead of you or behind you
- On the subject of throwing while walking
- On the subject of throwing the Wakizashi or Kaiken
- Important teachings regarding throwing Shuriken at night

- Secret Shuriken throwing technique

There are two lessons within Secret Shuriken throwing techniques

　　　The above lessons are all part of this license. This license is granted to a person who has done intense training and has therefore been granted license in all throwing techniques and Kuden, oral transmissions, of this school. This is what the holder of this Mokuroku represents.

(Signed)

Fujiwara Naritada
Fujiwara Yoshitoki

Chishin Ryu Shuriken Mokuroku
Chishin School: A Catalogue of Shuriken Lessons and Techniques

知新流手裏剣目録

一 手裏剣離之事
一 手裏剣軽重之事
一 同長短之事
一 同手之内之事
一 同足踏之事
一 打出目付之事
一 指屈伸之事
一 上下打之事
右八ケ条立打也

一 居打之事
一 左右打之事
一 二本打之事
一 三本打之事
一 四本打之事
一 三間打之事
手裏剣留打様
一 風切
右八ケ条也

一 夜打様
一 懐剣
一 腰刀
右三ケ条者免許之伝也

右之条々令相伝畢猶於鍛錬修行有
之者免許之伝口打可令相伝者也仍
目録如件

大和郡山之住士
飯嶋市兵衛

同
飯嶋源太左衛門

同
日置金左衛門

尾州之浪士
浅野伝右衛門

同国之住士
丹羽織江

Chishin Ryu Shuriken Mokuroku
Chishin School:
A Catalogue of Shuriken Lessons and Techniques

- How to Release the Shuriken
- The Proper Range of Weight for Shuriken
- The Proper Range of Length for Shuriken
- The Proper Way to Step When Throwing Shuriken
- Where to Look When Throwing
- How to Apply Proper Pressure With Your Fingers
- How to Strike Targets Above and Below You

The above 8 topics deal with throwing while standing

- Throwing While Seated
- How to Throw Left or Right
- Throwing Two Shuriken
- Throwing Three Shuriken
- Throwing Four Shuriken
- Striking a Target Three Intervals Away
- How to Throw Shuriken at a Target
- Cutting the Wind

These are 8 topics

- Throwing at Night
- Kaiken (A small knife in a scabbard kept inside the shirt.)
- Koshi-gatana ("Waist Katana" another word for Katana.)

The above three topics are taught to those who receive Menkyo, full transmission.

The certificate establishes that the holder received full transmission of all teachings of this school and has trained diligently. Granting of this certificate means that you may pass on these teachings as long as the person you instruct is an appropriate student.

Order of Succession:

Founder of the Chishin School
A Samurai of Yamatokori (In Nara Prefecture)
Ishima Ichihyoe
↓
A Samurai of Yamatokori
Ishima Genta Saemon
↓
A Samurai of Yamatokori
Heki Kinsaemon
↓
A Ronin of Oshu
Asano Denuemon
↓
A Samurai of Oshu
Tanba Orie

Granted on an auspicious day in August the 6th year of Horeki
1756

知新流手裏剣免許

当流手裏剣者知新流剣術之内抜出一流狄心不浅多年出精稽古之仮神妙之至候依

之夜之打形並懐剣腰刀打形不残令伝授候爲此上無怠慢工夫鍛錬可為専一候向後

狄心之當於有之者戯事心得不申様以固可有之指南候仍而免許如件

Chishin Ryu Shuriken Menkyo
Chishin School of Shuriken : License of Transmission

The teachings of this branch of Chishin Shuriken are taken from the curriculum of the Chishin School. After many years of dedicated training, without interruption, the bearer of this certificate achieved enlightenment to the inner mysteries of this school. Thus you have been taught every aspect of:

Throwing at Night
Throwing the Kaiken and Koshigatana

It is expected you will devote yourself entirely to forging your skill diligently and develop your own Kufu, or knacks and small improvements, to further advance your abilities.

You have been granted full authority to pass on these teachings and grant licenses, however you must be sure of the character of those you teach. If you judge them to be deceptive in character you should not instruct them. These are the conditions of this license.

Order of Succession:

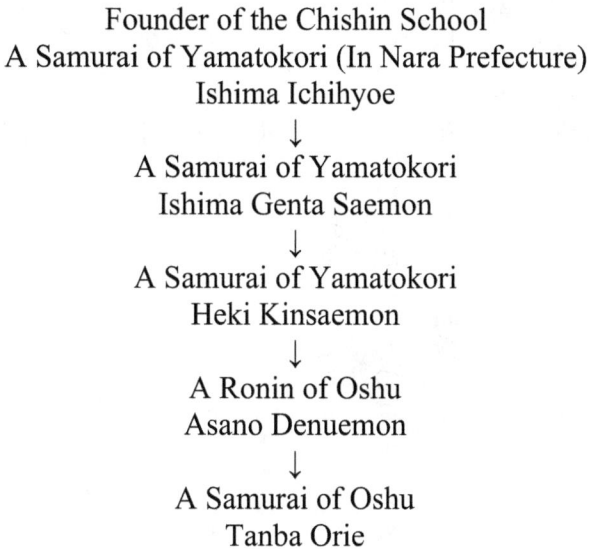

Founder of the Chishin School
A Samurai of Yamatokori (In Nara Prefecture)
Ishima Ichihyoe
↓
A Samurai of Yamatokori
Ishima Genta Saemon
↓
A Samurai of Yamatokori
Heki Kinsaemon
↓
A Ronin of Oshu
Asano Denuemon
↓
A Samurai of Oshu
Tanba Orie

Granted on an auspicious day in August the 6th year of Horeki
1756

印可伝授書

知新流手裏劍と云ふは強きに限らす弱きを厭はず兎に角早く打ち出し當るを専

一とする也　先の柄に手を掛ると見たら直ちに打ち出すなり　目録の内に打出

し目付と有るは是也　手裏劍を初めて教ゆるに先つ手前の右足を先の目當て先

きえ向けて踏み打つなり　打出す劍と足と一所に打出す事也足踏み専一也

足踏みそまつに心得ては夜の打様に不当　桃燈又は行燈杯有之場所にて打つ時

は光をおゝて打つ事也　拟劍を強くきかせんと思はゞ劍を柔かに持つて振り打

ちに打たは劍當りきくなり　拟又打劍の右より立つは離れに指先のきく故也

劍の左より立つは手のひらにて打つ故に押しつけ離れる故なり　劍の上より立

つは離れをおしむ故也　手離れをおしまぬ様に心得打つ事専一なり

笠かむり打つには間近き場所にて打つ心得よろし　手裏劍留に打ち通すもよろ

し　水にぬれたる劍は随分やわかに持ち打つなり　又　幼年の者に大劍を打た

するには肩えかけて打たする事　長き物を打つには中程のつり合を考え打つ事

なり

Chishin Ryu Inka Denjusho
Chishin School:
Certificate of Transmission and Authorization to Teach

The philosophy of Chishin School Shuriken is not focused on throwing hard or soft, rather it emphasizes throwing quickly and striking your target accurately. You should throw as soon as your opponent places his hand on the handle of his Katana. This is what is written in the *Where to Look When Throwing* chapter of the *Catalogue of Techniques of the Chishin School.*

The first lesson in our school of Shuriken is to aim at your opponent by using your right foot. Step forward with your right foot and throw the moment it contacts the ground. You should focus your training on throwing the moment your foot contacts the ground. If you are careless about how you step you will be unable to hit targets at night.

At night, when throwing a Shuriken at a person holding a paper lantern, block the source of light and throw.

If you want the blade to strike hard, hold the Shuriken lightly as you raise your hand and throw. This will cause your blade to strike hard.

If, after throwing, the end of your Shuriken is leaning to the right, the way your fingertips apply pressure when releasing is the problem. If the end of your Shuriken is leaning to the left, the way you apply pressure with your palm when releasing is the problem. If the end of the Shuriken is pointed up, you are releasing the Shuriken too late. You should devote yourself to finding the ideal way to release the Shuriken.

When throwing at a person wearing a straw hat remember you will need to be closer than you think. This also applies to throwing Shuriken at a target board.

If your Shuriken are wet, hold them as lightly as possible when throwing.

When teaching young people how to throw Daiken, large Shuriken, focus only on throwing with one arm.

When throwing a Shuriken at something long, aim somewhere in the middle of the target.

目録の内に四本打といふ有り　四本の劒を紙にて状を封ずる様に巻き劒はうち

違いに包み其儘打ては二本の劒はたしかに立つ事なり　又はひとえの袋に入れ

て打つ事なり　袱紗などに包み結ぶは悪るし　結び目のあらば打ちがたし　能

くく心得べし

劒がしらの下り立つは押付のきく故也　横平に當るは大指のすべる故也　又り

きみの有る故もあり　是れは能く見合せ直す事なり

懐劔手の内は両様におしえてよろし引上げの手の下らぬ様におしえる事也

懐劔長さは六寸より九寸五分迄　扨小劒にて板金を通すは劒のきく事を見るに

あらず手離れすなをに離るゝ事を見る為なり　劒のひずみて立つものは心能く

ぬけざる事なり　懐劔のさやを下え抜き打つ事なり

扨　手裏劔修行の人の心得と申すは常々客に参り候はゞ　先づ茶を出し候はゞ

茶碗を右の膝元に置く事　煙草盆を出し候はゞ随分と手近かに引付け置く事な

88

In the Mokuroku, or Catalogue scroll, there is a technique called Yon-hon Uchi, throwing four Shuriken at the same time. Wrap four Shuriken inside a piece of paper. Since two of the four Shuriken are facing the opposite directions when you wrap them up, at least two points will strike. You can also place them inside a bag and throw.

Wrapping them up with a Fukusa, a small silk cloth commonly used for cleaning tea ceremony implements, is not a good idea. The knot you tie will adversely affect your throw.

If the Shuriken is hanging down after you throw, then you have applied too much pressure. If the Shuriken is leaning either to the left or right after striking the target, it is because your thumb is slipping when you throw. When you see these kinds of errors, you should immediately correct your technique.

Our school teaches that the Kaiken, knife kept in the breast pocket, is handled the same way as a Shuriken, you shouldn't allow the hand you raise to throw to drop too low. A Kaiken should be between 6 ~ 9.5 Sun, or 18 ~ 29 centimeters, long.

Translator's Note:
Kaiken or "Breast Pocket Knife" was a common weapon carried by travelers. Carrying swords was illegal in the Edo Era unless you were a Samurai, however small knives were allowed. That being said the definition of "small" became rather elastic and travelers began carrying rather large "pocketknives."

You may have heard of throwing a Ko-ken, small sword, through a sheet of metal. However, it is not about skillfully throwing a blade, instead it is about making it appear as if you have thrown a blade when you have not.

If a blade bends when it strikes, it means you have completely lost your focus. This school teaches to pull the scabbard down off your Kaiken before you throw.

On another note, I would like to give some advice to those going on Shuriken Shugyo, travelling around and learning from different schools. Please behave like a normal guest. If they offer you tea when you arrive, place the teacup beside your right knee. If an ashtray is offered, be sure to pull it very close to you. While all the above applies to Shuriken, it also applies to Sansu, the Japanese folding fan.

り　手裏劒に限らず扇子にても右の心得よろし

飛龍劒　腰刀は抜き離し下えさげむねを平手にて持ちそえ　つる〳〵と足をは

こび間合を見て打ち離す事也　打ちかゝるに振り上げ打ち出す節しばらく刀を

引き出して離すことなり　打ちかゝる離れの節柄を下え引き下げ離すことなり

腰刀長さ一尺二三寸より七八寸よろし　長き腰刀は不好　目當は土俵にて打

たする事なれ共　近年は目當板にて打たする事になりたり　然し飛び返えり候

えばあやまち有る事故用心すべし

扨　門人にて無之人手裏劒所望之者有之候はゞ　初め五本随分やわらかに打ち

て見せる事なり　次に五本は常々稽古之通りに打ち見せ　三度目に板金を打つ

心にて打つ事也　三度之外打つ事なかれ　他人に打ち見せるには居打立打二本

打つ事　懐劒は二三本打ち見せる也　何れも沢山に打つ事なかれ　小劒は四品

の外かたく打つべからず

Hiryuken, Flying Dragon Sword. Draw your Koshi-gatana and hold it pointed down with your left hand on the handle and the palm of your right hand on the back of the sword. Press in on your enemy with smooth footsteps and then throw your sword like a Shuriken. When you are ready to throw bring the sword up into Jodan Kamae and, the moment you release, pull down on the handle. The best length for a Koshi-gatana is from 1 Shaku and 2 or 3 Sun to 7 or 8 Sun, 36 ~ 39 centimeters or 51 ~ 54 centimeters. Long Koshi Gatana are not good.

Translator's Note: These illustrations from *Two Hundred Illustrations of Weapons* 武器䡎圖 by Kobayashi Yuken 1848 show the differences between common swords and knives.

打刀 腰刀 右手指

Uchi Gatana (Katana)

Koshi Gatana (Katana)

Medezashi (Utility knife /Shuriken)

Samurai practicing Shuriken from *Illustrations of Learning* 写真学筆 by Maki Bokusen 牧墨僊 (1775－1824). The man on the bottom left is holding a Metezashi.

When a member of this school is demonstrating Shuriken throwing to a person not in the school, first throw five Shuriken in a gentle manner. Next, throw another set of five Shuriken as you normally would during training. For the final set, throw with power as if you are trying to pierce a metal plate. Do not throw more than three sets.

When showing others throw two times from a seated position and two times from a standing position. When demonstrating throwing the Kaiken, knife kept in the breast of the shirt, only throw two or three times. It is not necessary to demonstrate more than that. Never show anything more than the four basic throws with the small blades.

Translator's Note:
Dimensions of a Chishin School Shuriken:

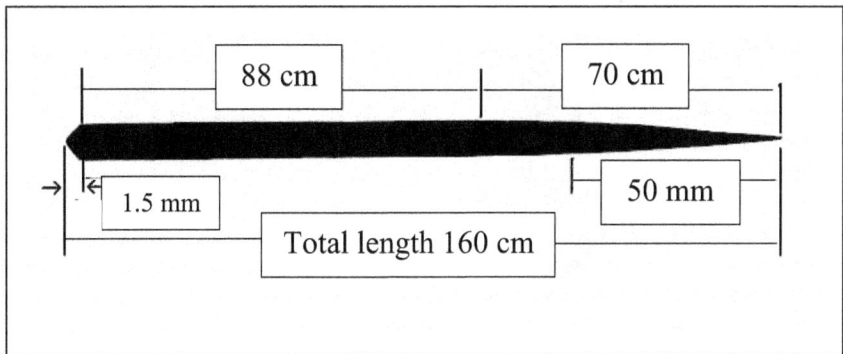

88 cm   70 cm

1.5 mm   50 mm

Total length 160 cm

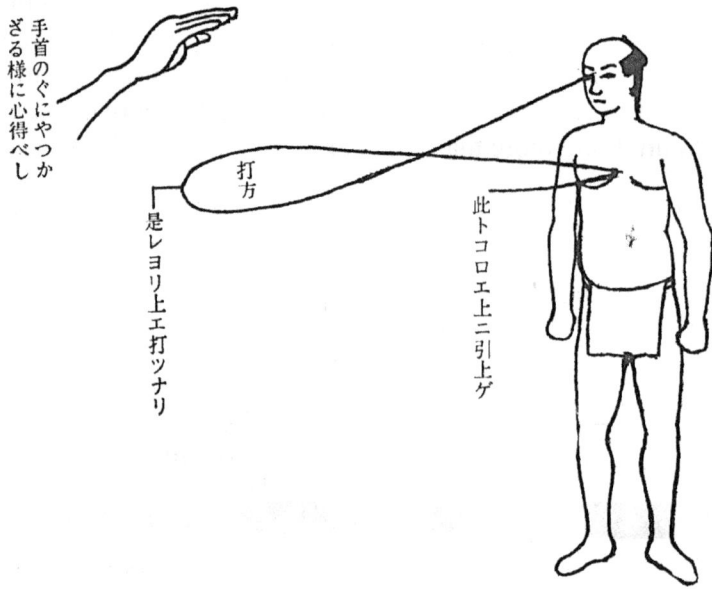

抜　劔を拵うるには　小劔は重さ二十五匁より三十五匁迄　長さ五寸より四五分迄二三分通用なり夫れも人々物好み次第なり

此トコロエ上ニ引上ゲ

打方

是レヨリ上エ打ツナリ

手首のぐにやつかざる様に心得べし

It is important to not allow your wrist to bend.

Direction of throw

You should throw before this point.

Raise above this point.

擬劍を拆うるには　小劍は重さ二十五匁より三十五匁迄　長さ五寸より四五

分迄二三分通用なり夫れも人々物好み次第なり

If you are going to make your own Shuriken, small blades should weigh between 25 Monme and a maximum of 35 Monme, 94 ~ 131.25 grams. The length should be between 5 Sun and 4 or 5 Bun, 16 ~ 16.5 centimeters. The weight of most Shuriken is within 2 or 3 Bun, 0.6 ~ 0.9 centimeters, however it is up to each person to decide for themselves.

Translator's Note:
The old measurement varied considerably over time as well as by location. I have used these measurements:
One 寸 Sun is 3.03 centimeters.
One Bun is .3 centimeters
One Monme is 3.75 grams

此の通り指の
直なる様に

足先を向の足先えむけて引上る手と一所に足を踏込み打つ也　打ちかくる手と

一所に踏込む事也

初め人に教ゆるに　五六尺の間合にて随分やわらかに打ちおしゆる事なり　追

々剣の立つに付いて　足首丈つゝ　しさり　次第〳〵にしされば本間えなる事

なり

Point the end of your foot at the tip of the opponent's foot. Raise your arm as you step and release as your foot contacts the ground. Throw the Shuriken and plant your foot at the same time.

The distance for beginners should be around 5 or 6 Shaku, 150 ~ 180 centimeters. Teach them how to throw gently. Later, after they become able to stick Shuriken in the target, increase the distance little by little, one foot-width at a time.

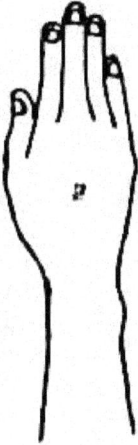

Your fingers should be kept straight like this.

稽古の節劔を取り落す事有り　直ちに拾ろいて打つ事悪し　落した劔にかまわ
ず手に有る劔を打ち切りしあとにて拾い打つ事なり　劔を手の内え能くなつけ
る様に指をつたい様に打つ事専一なり
劔を打つに我が手のひらの先え見ゆる様に心得打つ也　打ち出す時劔より足の
先え出る事悪るし　劔と足と一所に踏込む事なり
大指の離れぬ様に心得打つ事也　大指すべれば劔横ひらに當るなり　又劔の離
れをおしめばたてに立つなり　懐劔初めて教ゆるに手の内は定法なり　目付は
鼻より下を心がけ打つなり

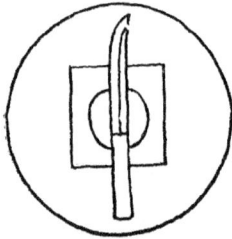

此の通りに当る様に不
立様に打たせる事也ケ
様教えれば則ち立なり
是れ秘伝なり

98

When training you may accidentally drop a Shuriken on the ground. However it is bad form to immediately pick it up and throw it. Don't concern yourself with the Shuriken you dropped and instead continue throwing the ones remaining in your hand or belt. After finishing that pick up the one you dropped and throw it. It is important for you to learn how to handle the Shuriken. You need to focus completely on training your fingers develop a sense of how to hold the Shuriken so you can throw them comfortably.

Concentrate on looking past the ends of your finger when you throw Shuriken. When throwing, allowing your foot to move before your blade is a bad habit. The blade and your foot should move in unison.

Be sure to not allow your thumb to lose contact as you throw. If your thumb slips it will cause the blade to stick in the target leaning to one side or the other.

If you release the Shuriken too late it will stick in the target vertically.

There is a standard way to introduce how to handle and throw the Kaiken. You should focus on throwing at a point below the nose.

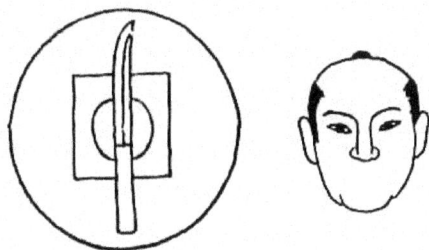

此の通りに当たる様に不立様に打たせる事也ケ様教え
れば則ち立なり是秘伝なり

This diagram illustrates how you should throw and how not to throw. If taught according to these principles, you will immediately be able to throw accurately. This is a secret teaching.

懐剱と腰刀は離れのおしみて放つ心也

懐剱之目付は鼻より下を打つ心也　懐剱は胸より下らぬ様に打つなり

剱に十分一之剱というは八十目の剱一本に八匁の剱二本是れを八匁剱を打つ

次に八十目の剱を打ち　又八匁の剱を打つケ様に入違いに打たすれば大剱にて

も小剱にても打ち覚える為なり

丸き物を打つには指三本かけて打つ也左之通り

When throwing Kaiken or Koshi Gatana you should hold on to the handle a bit longer before releasing. When releasing the Kaiken you should focus your aim at a point below the nose. However you should never allow the Kaiken to strike below the chest.

The training method "One in Ten Shuriken" refers to throwing 1 or 2 Shuriken weighing 8 Monme, or 30 grams, for every 8 ~ 10 regular Shuriken. After throwing 1 or 2 of the 8 Monme Shuriken, throw 8 ~ 10 regular Shuriken and then mix in the 8 Monme Shuriken. By switching the weights around you can get used to throwing both Daiken, long Shuriken, and Shoken, small Shuriken.

When throwing objects with round handles you should hold and throw with three fingers. This is shown in the illustration below.

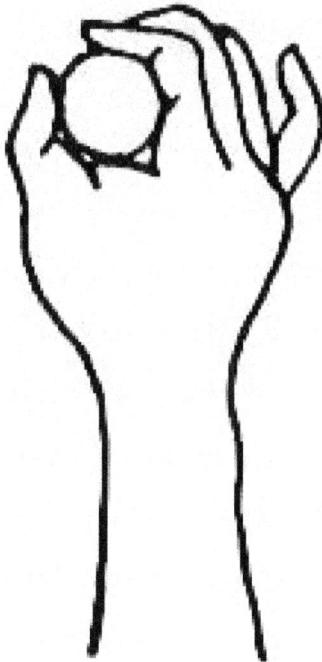

居打は打出す節いしきを少し上げて度々に打つ事也

手裏劔留は先の左より返しこみ打つ事也　常々稽古に足踏み第一也　初めに教ゆる節足は目當の通りえ右足を踏み振り上げ打ちかゝる時足を一所に踏み出す

足の踏み出す事劔より早く出るは悪し

上下左右之乱劔の者えは目當を見はる事悪し　劔取る節手本を見て振り上げると一諸に目當てを見て打つか　又は手前の足先を見て振上げる迄目當を見る事悪し

When throwing from a seated position, focus your aim slightly higher than your actual target. Practice this repeatedly.

Shuriken-tome, Shuriken Stop, is when you throw across your left side.

The most important lesson to train repeatedly is how to step. This should be the first thing you teach beginners. Your right foot should be aimed at your target. Then, step forward as your arm begins to rise. Your throw and your foot making contact with the ground should be simultaneous. If your foot lands before you throw, you are doing it wrong.

People that end up with wild throws, as in the Shuriken stuck in the target facing upward, hanging downward, leaning left or right, are doing a bad job of picking their target. Such a person may be looking at their hands when drawing the Shuriken and then focusing on the target as they raise their arm. Another possibility is they are focusing on the leading foot as they raise their arm to throw. This is a poor way to aim your throw.

常々足踏みは左の通りなり

○

打出す時此処え
踏み出すなり

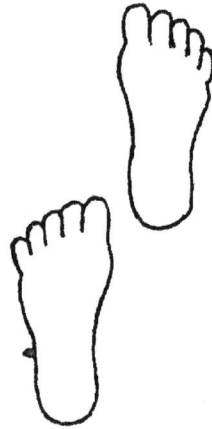

右の通り足踏みは左足の大指の頭通りえ　右足のきびすを踏出す時右足を二ツ
丈踏み出す也　又右足を踏み付けて打つもよろし　左足はきびすを踏付けぬ様
に心得専一也　万事足は軽く踏む事なり
遠間はさかに劔を取り打ち出す時指先にて少し押える様にあしらいて離す事な

You should study the illustration below carefully; it shows how to plant your feet. This is how you should be stepping when you throw.

打出す時此処え

踏み出すなり

○

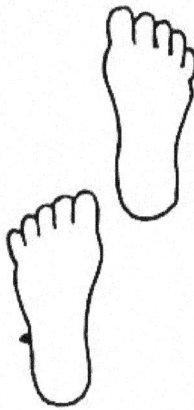

As you step, your right foot should pass over the big toe of your left foot. When stepping, your right heel should travel about two-foot spans. As I said before you should throw as your right foot contacts the ground. Another important point is not allowing the heel of your left foot to be planted on the ground, keep it off the ground. Always be sure to step lightly when moving.

When throwing a long distance you will be holding the Shuriken reversed in your hand. As you release, press you're your fingertips.

り

稽古五本劔を左の腰通りえ下げ　一本づゝ取り打つ事也　腰通りと云うは劔打

懸り直ちに刀の柄に手のかかり申す為めに持ち覚ゆるためなり　五本剣は劔と

心得ず　一本打出し直ちに柄に手をかけ申す心なり　刀の鯉口を持つ心なり

当り左右え乱劔を直す事　巾二三分長さ目当板の丈に白紙をたち板えたて張り

打すれば左右はづれ直る事なり

甘シャウ剣と云う有リ　是は一子相伝同様の事なれば印可つかはし候とも此カ

ンシャウ剣は伝授無用也

目当四寸に五寸と定むる事は元祖飯嶋氏竹村与右衛え打見せ候節　目当は何程

に致し稽古致し候やと尋ね候時四五寸の目当と申すに付き四寸に五寸と定むる

なり

106

When training, keep five Shuriken in your belt on your right side to draw with your left hand. This is known as Koshi Dori, Waist Drawing. Draw them one at a time and throw. The reason to train throwing with Shuriken in your belt on your right side is to train your left hand to shift to the handle of your Katana after each throw. Though you have five Shuriken in your belt by your side, the purpose of this exercise is to train your hand to immediately grab the handle of your sword after each throw. Focus on grabbing the Koi-guchi, the top of the scabbard below the sword guard.

The next topic is how to correct wild throws that lean to the left or right. Prepare a piece of white paper 2 or 3 Bun, 6 ~ 9 millimeters, wide and as long as a Shuriken target board. Throwing at this will correct the problem of Shuriken leaning to the right or left after striking the target.

Within the teachings of the Chishin School there is a technique called Ama Sho Ken (meaning unknown.) This teaching is only passed on to a single student and thus is not part of this certification and will not be taught to you.

The founder of this school, Ijima Sensei, who was a student of Takemura Youemon, once asked him, "What are the proper dimensions for a Shuriken target?" In response Takemura answered, "I would say 4 or 5 Sun, 12 ~15 centimeters, is best." Therefore we have decided to train with both 12 and 15 centimeter sized targets.

此印可之一巻者手裏剣伝授之薀奥也然処貴殿之執心他勝殊修行不怠故授与之旱

後門人に伝えむとあらば必ず其人の器量を計りて可伝妄りに不可伝仍而奥書

如件

大和郡山之住士

流祖　飯嶋　市　兵　衛

同

飯嶋源太左衛門

同

日置　金左衛門

尾州之浪士

浅野　伝右衛門

同国之住士

丹　羽　織　江

108

You are receiving this certification scroll authenticating that you have learned the inner mysteries of this art. Your dedication to training has exceeded that of others and you never were lax in your Shugyo, intense learning. You are now authorized to teach the Chishin School Shuriken arts to students, however if you sense a person is not of good character, absolutely do not teach them. That is a condition of this certification.

Order of Succession:

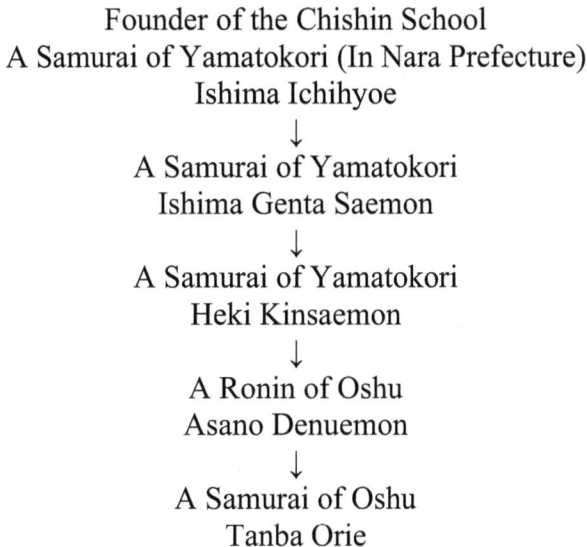

Founder of the Chishin School
A Samurai of Yamatokori (In Nara Prefecture)
Ishima Ichihyoe
↓
A Samurai of Yamatokori
Ishima Genta Saemon
↓
A Samurai of Yamatokori
Heki Kinsaemon
↓
A Ronin of Oshu
Asano Denuemon
↓
A Samurai of Oshu
Tanba Orie

Granted on an auspicious day in August the 6th year of Horeki
1756

手裏剣の長さは、流派により、人によって一定した寸法はないが、五寸または三寸を基本定寸としたものである。五寸は、五行を表わし、空風火水地の人体に討ち込むという意を現わしたもので、また、三寸は、日月星の三光、すなわち、朝の明星と空の日月の三光を以て、不絶剣という意である。

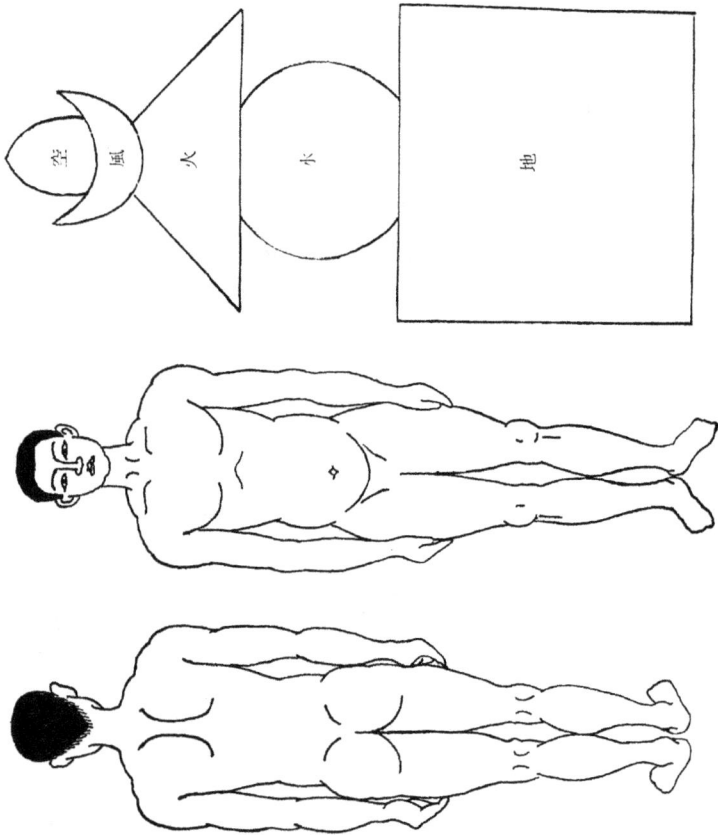

次に各流派特定の手裏剣の型態を示す。

各流手裏剣の形態（イ）

四五

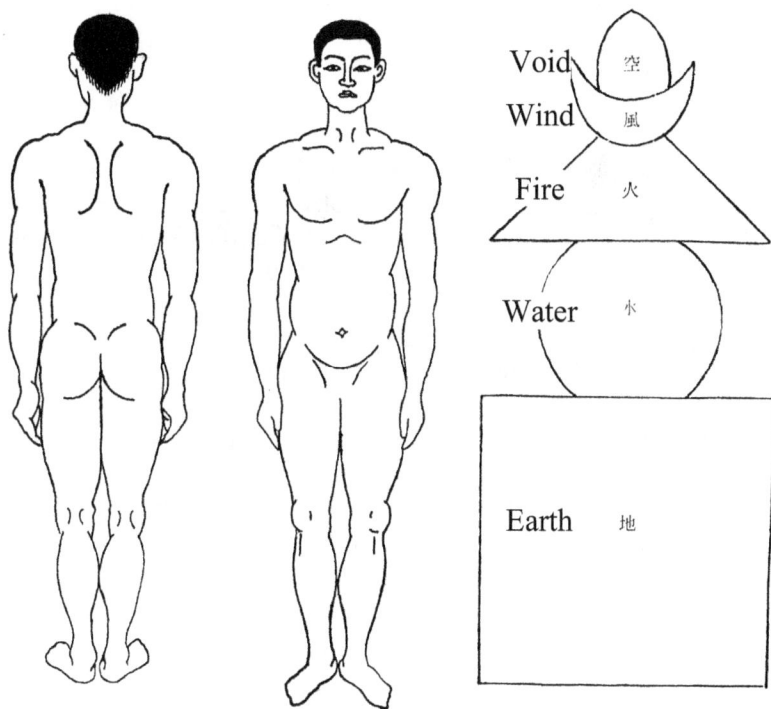

Void 空
Wind 風
Fire 火
Water 水
Earth 地

Translator's Note: These are known as *Gorinto* 五輪塔, Five Ringed Pagodas. They are typically made of stone.

手裏剣の長さは、流派により、人によって一定した寸法はないが、五寸また
は三寸を基本定寸としたものである。五寸は、五行を表わし、空風火水地の人
体に討ち込むという意を現わしたもので、また、三寸は、日月星の三光、すな
わち、朝の明星と空の日月の二光を以て、不絶剣という意である。

Shuriken can vary in length according to the Ryuha, or school. While are no prescribed dimensions, 5 Sun (15 cm) or 3 Sun (9 cm) are well-established dimensions.

The 5 Sun length reflects the Five Elements: Sky, Wind, Fire, Water and Earth. These five elements correspond to the places you strike on the human body.

The 3 Sun length refers to the three celestial bodies, the sun, the moon and the stars. In other words, Venus, the morning star, along with the two other sources of light in the sky, the sun and the moon. It implies the Fuzetsuken, the Immutable Sword.

Translator's Note:

The next two illustrations are from Fujita Seiko's 1958 拳法極意 當身殺活法明解 *A Clear Explanation of Striking Points and Resuscitation Points.* They contain more information on how the Gorinto relates to the human body.

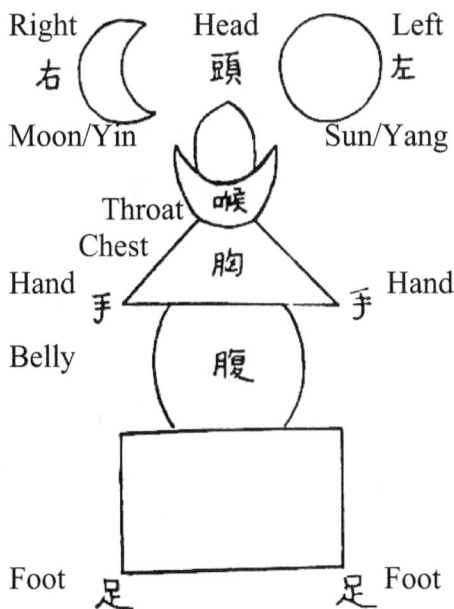

古い伝書には、直接人体を繪がかず、五輪の塔を以て五体に擬へ、日月陰陽を以て左右を示し、頭部、咽喉、胸、手、腹、足に相当する部分〈に只大体の名称を付して急所の位置を示し口伝を以て伝受する方式をとつてゐた。

Right — Head — Left
右 — 頭 — 左
Moon/Yin — — Sun/Yang
Throat 喉
Chest
Hand 胸 Hand
手 — 手
Belly 腹
Foot 足 — 足 Foot

In old Densho, the human body was usually not drawn. Instead a Gorinto, Five Storied Pagoda, was used to represent the five parts of the body. Yin and Yang were depicted with the moon representing your left and the and the sun representing your right. The illustration thus generally displays where the head, throat, chest, hands, belly and feet are located. The Kyosho, or striking points, were all transmitted orally. This is how information was passed down.

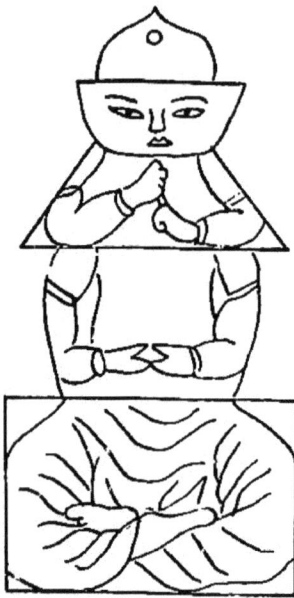

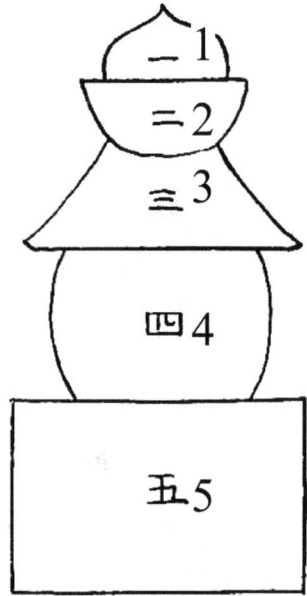

五体を象る五輪塔
Gotai wo Katadoru Gorinto
How the body is represented in a five-storied pagoda.

各流手裏剣の形態

Shuriken Measurements for Each School Part One

心月流　　　　　　　長サ七寸　径二分三厘

　　　　　　　　　　長サ四寸七分　径二分七厘

伊豆流

　　　　　　　　　　長サ五寸八分　径二分五厘

伊豆流

願立流　上遠野流　　長サ六寸八分　径三分

白井流　　　　　　　長サ七寸　径二分

根岸流　　　　　　　長サ三寸　径二分五厘

根岸流　　　　　　　長サ四寸五分　径五分

願立流 上遠野流 長廿六寸八分 径一分	伊豆流 長廿五寸八分 径一分五厘	伊豆流 長廿四寸七分 径一分七厘	心月流 長廿七寸 径二分三厘
Ganryu School Kadono School Length : 6 Sun 8 Bun 20.4 cm Diameter : 2 Bun 6 mm	Izu School Length : 5 Sun 8 Bun 17.4 cm Diameter : 1 Bun 5 Ri 4.5 mm	Izu School Length : 4 Sun 7 Bun 14.1 cm Diameter : 1 Bun 7 Ri 4.1 mm	Shingetsu School Length: 7 Sun 21 cm Diameter: 2 Bun 3 Ri 6.9 mm

117

根岸流

根岸流

白井流

長廿四寸五分　径五分

長廿三寸　径二分五厘

長廿七寸　径二分

Negishi School	Negishi School	Shirai School
Length :	Length :	Length :
4 Sun 5 Bun	3 Sun	7 Sun
13.5 cm	9 cm	21 cm
Diameter :	Diameter :	Diameter :
5 Bun	2 Bun 5 Ri	2 Bun
1.5 cm	7.5 mm	6 mm

長サ四寸八分　径三分五厘

長サ六寸　径三分

諸賞流　　狐伝流

長サ四寸五分　径三分

長サ四寸六分　元径三分五厘

盂　淵　流

長サ三寸　ないし　三寸五分

長サ五寸五分　三分角

長サ四寸四分五厘　元一分五厘角

119

Koden School Shosho School Length : 4 Sun 6 Bun 13.8 cm Diameter : 3 Bun 5 Ri 1.05 cm	Koden School Shosho School Length : 4 Sun 5 Bun 13.5 cm Diameter : 2 Bun 6 mm	Negishi School Length : 6 Sun 18 cm Diameter : 3 Bun 9 mm	Negishi School Length : 4 Sun 8 Bun 14.4 cm Diameter : 2 Bun 5 Ri 7.5 mm

		孟淵流
長廿四寸四分五厘　元一分五厘角	長廿五寸五分　三分角	長廿三寸　ないし　三寸五分

Moen School	Moen School	Moen School
Length :	Length :	Length :
4 Sun 4 Bun 4 Ri	5 Sun 5 Bun	From 3 to 3.5 Sun
13.5 cm	16.5 cm	9 ~ 10.5 cm
The thickest part at	Diameter across	
the cross-section is	square bottom :	
1 Bun 5 Ri	3 Bun	
4.5 mm	9 mm	

長サ五寸九分

香取神刀流

長サ五寸三分 径三分

丹知新
羽流流
流

長サ五寸三分五厘 三分五厘

長サ四寸八分 幅三分三厘 厚三一分六厘

津　川　流

長サ六寸三分 幅五分三厘 厚サ一分

伊　賀　流

伊　賀　流

122

知新流　丹羽流

香取神刀流

長サ四寸八分　幅二分一厘　厚サ一分六厘

長サ五寸二分五厘　三分五厘

長サ五寸二分　径二分

長サ五寸九分

Chishin School Tanba School Length : 4 Sun 8 Bun 14.4 cm Width : 3 Bun 3 Ri 10.1 mm Thickness : 1 Bun 6 Ri 4.8 mm	Chishin School Tanba School Length : 4 Sun 8 Bun 14.4 cm Diameter : 3 Bun 5 Ri 4.5 mm	Katori Shinto School Length : 5 Sun 2 Bun 15.6 cm Diamter : 2 Bun 6 mm	Moen School Length : 5 Sun 9 Bun 17.7 cm

伊賀流	伊賀流	津川流 長廿六寸二分　幅五分三厘　厚寸一分
Iga School	Iga School	Tsukawa School Length : 6 Sun 3 Bun 18.9 cm Width: 5 Bun 3 Ri 1.59 cm Thickness : 1 Bun 3 mm

甲賀流　　　　　長サ七寸　元幅七分

諸賞流　　　　　長サ七寸　元幅七分五厘

竹村流　　　　　長サ八寸　元幅八分五厘　棟厚サ五分

堤宝山流

125

竹村流	諸賞流	甲賀流
長サ八寸　元幅八分五厘　棟厚サ五分	長サ七寸　元幅七分五厘	長サ七寸　元幅七分
Takemura School Length : 8 Sun 24 cm Width at the base : 8 Bun 5 Ri 2.55 cm Thickness of the back of the Shuriken : 5 Bun 1.5 cm	Shosho School Length : 7 Sun 21 cm Width at the base : 7 Bun 5 Ri 2.25 cm	Koga School Length : 7 Sun 21 cm Width at the base : 7 Bun 2.1 cm

堤
宝
山
流

Tsutsumi
Hozan School

Takemura
School

菱鉄

長さ二寸、幅一寸二分、厚さ一寸くらいの稜角をもった菱形の鉄塊の中央に穴を穿った物。

これをいくつでも紐に通して持って行き、敵に投げつける手裏剣の一種。

鹿子玉

（流派により、弾き玉、隠し目潰し等の別名あり）

これは鏡新明知流の鹿子玉。丸さ五分、針三分。

敵の顔面、目、喉等をねらって、投げつけたり、指先で弾きつける。

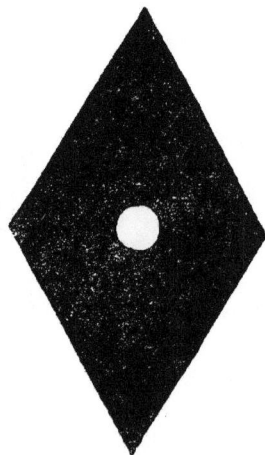

128

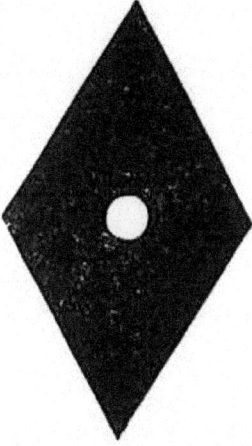	Hishi Tetsu Length: 2 Sun 6 cm Width : 1 Sun 2 Bun 3.6 cm Thickness : Around 1 Sun 3 cm This Shuriken is a piece of iron shaped like a Hishi, water chestnut with a hole drilled in the middle. You can carry as many of these as you want by threading a piece of rope through the hole. This is a type of Shuriken that can be thrown at your enemy.
	Kanoko (Meaning unknown, possibly based on a flower) Depending on the school of Shuriken this can also be called Spinnin Ball and Concealed Blinding Shuriken. This is a Kyoshin Meichi School 鏡新明智 (知)流 Kanoko Shuriken. The diameter of the circle is 5 Bun, 1.5 cm, with the points being 3 Bun, 9 mm in length. Aim for the eyes or throat of your enemy while throwing Kanoko Shuriken. When releasing, use your fingertips to make the Shuriken spin.

投箭

前の欧州大戦の時、仏軍では、投箭なるものを多数造って、これを飛行機の上から、独軍の密集部隊の上に投下して非常な効果を収めた。これは日本の手裏剣の一つが、巴里の兵器参考館に陳列されてあったのを見た仏国の一技術将校がそれによってヒントを得てこれを模造して試みたのが、はじめであった。

この投箭は、長さ十二センチ、中径八ミリ、重量十五グラムで、一機に三千五百から五千くらいを携行投下したのであるが、発射もせず機関銃射と同様の効果を挙げた。

二千メートルの空から投下すれば、地上近く秒速百五十メートル、キロメートルで優に百メートルに達して、その貫通力は、馬上の人の肩から突入、臀部に抜け、さらに馬の胴を貫通したという。この投箭は後いられ、

英軍においても長さ十三センチとなし基部を投下するように工夫され、逆に英次いで独軍でも、やがて矢が自転しないが、仏軍を苦しめたがその使用もまもなく休戦近くなって、この第二欧州戦でも独軍はただちにこれを使用した。

Tosen : Flechette

In the first European war, the French military created a multitude of Tosen, flechettes, and loaded them on airplanes. They flew over concentrations of German soldiers and dropped them from above, inflicting horrific casualties. When I visited the military museum in Paris, I saw all the flechettes lined up. The French Military Academy had a Japanese Juji Shuriken, Cross-Shaped Shuriken on display, and used that as a model for their flechettes.

The French flechettes are about 12 centimeters long with a diameter of about 8 millimeters and weigh about 15 grams. One aircraft can drop between 3,500 and 5,000 of them. Though they don't need to be fired, the effect is similar to that of a machine gun. Dropping them from a height of 2,000 meters will enable them to reach a speed of 150 meters per second. Dropping them from 1,000 feet will enable them to reach a speed of 100 meters per second. The force in a flechette travelling at such a speed is considerable. If one were to strike the shoulder of a mounted rider, it would pass through him, out his buttocks and then travel completely through the abdomen of the horse he rode.

As a result of these attacks, the British military developed their own design, followed by the Germans who developed a 13 centimeter version. The German flechette had a twist at the end which caused the metal arrows to rotate by themselves after being dropped. In a surprising reversal, the German made flechettes caused a great deal of pain to the British and French forces. However, these flechettes were not developed until near the end of the war, so their production soon ended. However, at the beginning of the Second European War, the Germans again began employing them.

German	British	French

独

英

仏

各流手裏剣の形態

Shuriken Measurements for Each School

·

Part Two

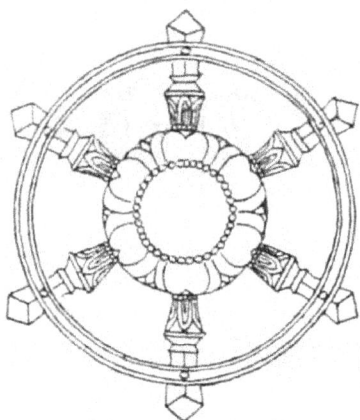

法輪
Horin

These are two examples of Horin, which represent the teachings of Buddha. They resemble the Dharmachakra, originally a wheel-like weapon used to destroy the evils of mankind.

法輪には、五輪宝、六輪宝、八輪宝等がある。その五輪宝、六輪宝、八輪宝が、五方、六方、八方等の手裏剣の造られる因をなしたものである。

法輪や、小楯流の五方、六方、八方等はそれである。また、その他の手裏剣も多くはその信仰観念に基づいて製作されたもので、羯摩が十字手裏剣である。

法輪は古代印度の武器の一種で、通常車輪の形をなし、輪辺には鋭い剣双をつけている。

これは投げつけて敵を倒すに用いたもので、この法輪はもと転輪聖王が仏法守護のために持した武器で、法輪は旋転運動をして大地の凸凹を平均し一切の障碍を破砕する功徳を有するものとしている。これにならって出来たのが車剣である。

独鈷

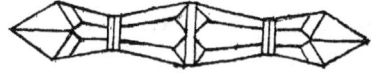

三鈷

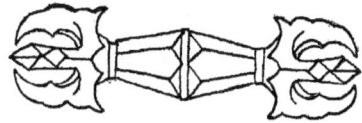

五鈷

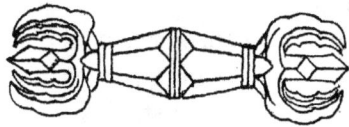

羯摩

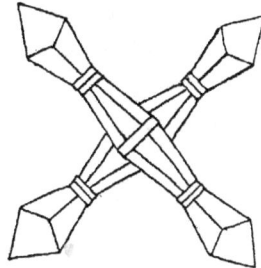

十字と卍字は、悪魔撲滅と招福除災の呪符に用いられる。

☆☆☆（星状）は、五行の金水火土木、五大（空風火水地）を表章したもので、五行五大成就、悪魔退散呪符である。また卍字は、萬徳の集まる吉祥の相象で、萬字の護符である。卍字には左右があり、右旋卍字・日・太陽を表章したものと、左旋卍字・月を表章したものである。左卍字を陰とし、右卍字を陽とする。

Horin Wheels

There are many kinds of Horin Wheels, with differing numbers of spokes. There are versions called Five Treasured Spokes, Six Treasured Spokes, Eight Treasured Spokes and so on. Similarly, there are Shuriken with 5 directional spikes, 6 directional spikes and 8 directional spikes, which reflect the Five Treasured Spokes, Six Treasured Spokes and Eight Treasured Spokes.

So, the Horin Wheels can have 5, 6 or 8 spokes and the Shuriken made in the Kobori School can have 5, 6 or 8 spikes. Further, other schools infuse their religious tenants into their Shuriken when making them. For example, the Juji, cross shaped, Shuriken represents Karma.

The Horin Wheel is a weapon that came from ancient India. It is shaped like a wheel with blades around the edges. These are Shuriken which were thrown into the opponent and stuck in, thereby toppling them. These weapons were originally created by an Indian Chakravartin ruler in order to protect Buddhism. In Indian religions the word chakravarti refers to an ideal universal ruler who controls all of India.

This Shuriken wheel has the power to grant the user the ability to use its spinning power to flatten uneven ground and, in one pass, obliterate any obstacles. The Shaken, Wheeled Shuriken, is based on this.

The cross shape in the Juji Shuriken serves as a talisman to eradicate evil spirits, defend against fire and disaster as well as bring luck.

The five pointed star represents the Five Elements : Wood, Fire, Earth, Metal and Water. It also represents the Five Great Forces :Air, Wind, Fire, Water and Earth and the Five Parts of the Body : Head, Neck, Torso, Hands and Feet. The joints serve as talismans against fire and misfortune.

The Manji 卍, which can also be written as "10,000 Kanji" represents a good luck symbol that gathers 10,000 virtues. There are two versions of the Manji, one rotating left the other right. The left rotating Manji 卐 is called Hidari Manji. The right rotating Manji is called Migi Manji 卍. The Hidari Manji represents Yang, light and the sun, while the Migi Manji represents Yin, dark and the moon.

羯磨 Katsuma	五鈷 Goko	三鈷 Sanko	独鈷 Dokko

三鈷、五鈷は、これを投げるというより、手に握り持って、敵を撃ち突きする、仏法守護のための武器の一種で、撃突武術用の陰拳、三節、他力、五輪砕、微塵等々の武器と同様のものである。

陰拳
荒木流 本覚克已流

三節
為我流

他力
本覚克已流

五輪砕
伊賀流

The Sanko, or Trident Vajra, and Goko, or Five Pronged Vajra, are not really meant to be thrown, rather they are gripped in the hand and used to hit or stab the enemy. The purpose of this weapon is to guard Buddhism from its attackers. It is similar to other striking weapons like Secret Fist, Three Arrows, Using the Power of Others, Breaking the Five Rings, Pulverize and so on.

Gorin Kudaki 五輪砕 Five Ring Breaker Both from the Iga School 伊賀流	*Tariki* 他力 Other Power Hongaku Kokki School 本覚克己流	*Sanbyaku* 三佰 Three Hundred Iga School 為我流	*Kage Ken* 陰拳 Secret Fist Araki School 荒木流 Hongaku Kokki School 本覚克己流

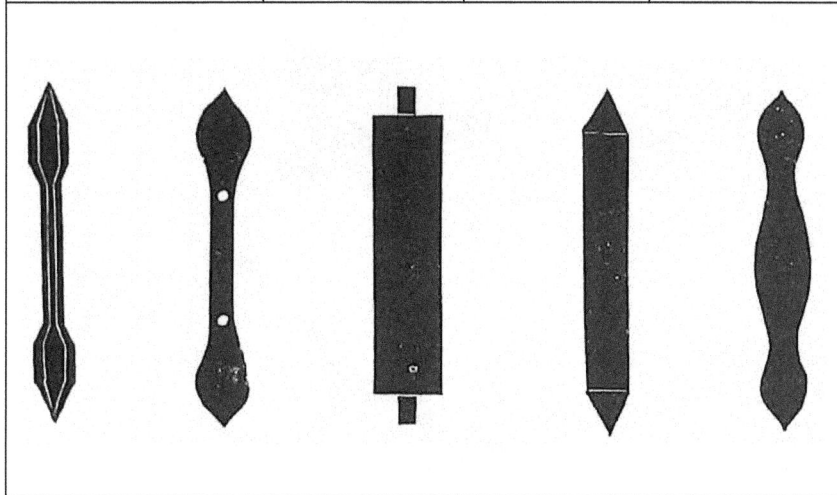

三光手裏剣

長サ中央円四分
一角剣長サ一寸三分

四方手裏剣

長サ三寸四分

五方手裏剣
一名星形手裏剣

長サ三寸

小堀流
六方手裏剣

長サ三寸
中央穴一寸

伊賀流
Iga School
Sanko Shuriken
Triple Light Shuriken
The center circle is 4 Bun, 1.2 cm in diameter. The angled blades are 1 Sun and 2 Bun, 3.6 cm long.

伊賀流
Iga School
Shiho Shuriken
Cardinal Directions Shuriken
Length : 3 Sun 4 Bun 10.2 cm

伊賀流
Iga School
Goho Shuriken
Five Pointed Shuriken
also known as Hoshjo or Star Shaped Shuriken.
Length : 3 Sun 9 cm

小堀流
Kobori School
Roppo Shuriken
Six Pointed Shuriken
Length : 3 Sun 9 cm
Diameter of opening at center 1 Sun, 3 cm.

甲賀流　伊賀流
六方手裏剣

長サ三寸

甲賀流　伊賀流
八方手裏剣

長サ三寸

寸

甲賀流　伊賀流
八方手裏剣

長サ四寸

小堀流
八方手裏剣

長サ三寸三分
中央穴六ノ寸

142

	甲賀流 Koga School 伊賀流 Iga School Roppo Shuriken Six Pointed Shuriken Length : 3 Sun 9 cm
	甲賀流 Koga School 伊賀流 Iga School Happo Shuriken Eight Pointed Shuriken Length : 3 Sun 9 cm The edges of this are sharpened.
	甲賀流 Koga School 伊賀流 Iga School Happo Shuriken Eight Pointed Shuriken Length : 4 Sun 12 cm
	小堀流 Kobori School Happo Shuriken Eight Pointed Shuriken Length : 3 Sun 2 Bun 9.6 cm The circle at the center is 1 Sun, 3 cm in diameter.

長サ三寸四分

十方手裏剣
甲賀流
伊賀流

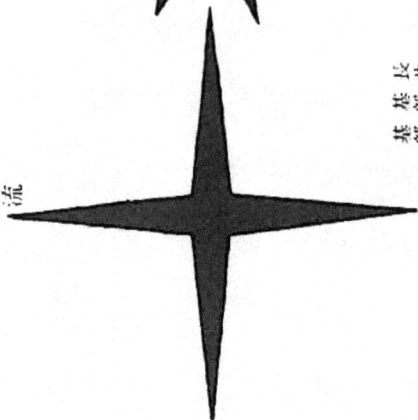

長サ四寸五分
基部幅四分
基部厚二分一分

十字手裏剣
柳生流
甲賀流
伊賀流

長サ三寸

十字手裏剣
柳生流

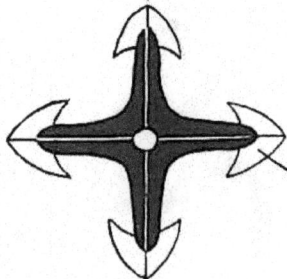

長サ四寸

み

十字手裏剣

144

甲賀流
Koga School
伊賀流
Iga School
Juji Shuriken
Ten Pointed Shuriken
Length : 3 Sun 4 Bun 10.2 cm

柳生流
Yagyu School
甲賀流/伊賀流
Koga School /Iga School
Juji Shuriken
Ten Pointed Shuriken
Length : 4 Sun 5 Bun 13.5 cm
Width at center : 4 Bun 1.2 cm
Center thickness : 1 Bun 3 mm

柳生流
Yagyu School
Juji Shuriken
Ten Pointed Shuriken
Length : 3 Sun 9 cm

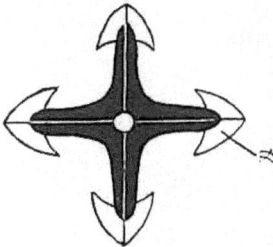

柳生流
Yagyu School
Juji Shuriken
Ten Pointed Shuriken
Length : 4 Sun 12 cm
The blades at the end are sharpened.

伊賀流甲賀流では
四方手裏剣という
狐伝流では
中剣という。

狐伝流では
中剣とも
糸巻剣とも
いう。

小堀流
万字手裏剣

十字手裏剣の変形で
あるが、正しくは
万字手裏剣という。

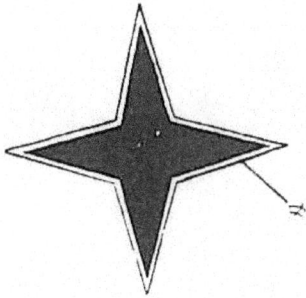	伊賀流・甲賀流・狐伝流・諸賞流 The Koga and Iga Schools both call this a Shiho, Four Pointed, Shuriken but the Koden and Shosho Schools call it a Shaken, Wheel Blade. It is 3 Sun, 9 cm, in length and the edges are sharpened.
	狐伝流・諸賞流 Both the Koden and Shosho schools call this either Shaken or Itomaki Ken, Wrapped String Shuriken. The length is 3 Sun, 9 cm, and the edges are sharpened.
	Kobori School 小堀流 Manji Shuriken 万字手裏剣 Length 2 Sun 3 Bun 6.9 cm
	This is a variation of the Juji Shuriken, however it is properly known as a Manji Shuriken. It is 3 Sun 2 Bun, 9.6 cm in length with a thickness of 1 Bun, 3 mm. The hole in the center is 2 Bun, 6 mm in diameter. The Tsume, or fingernails, of the Shuriken are sharpened.

法輪

九宮

窗口

四方十字

修所建道住洞流

	法輪 Horin Buddhist Wheel
	九字 Kuji Nine Seals
	篭目 Kagome Basket Weave
	新陰流 Shin Kage School 龍首剣 Ryushuken Dragon's Neck Blade 4 Sun 4 Bun 13.2 cm

四寸四分

狐伝流	新陰流
Koden School	Shin Kage School
諸賞流	Sanko 三光
Shosho School	4 Sun 4 Bun 13.2 cm

十字形手裏剣

二つの剣を一つに合わせ、開くと十字形手裏剣となるように出来た手裏剣で、バネの仕掛があって、たたむと一つになり、開くと十字形となってバネがかかるように出来ている。

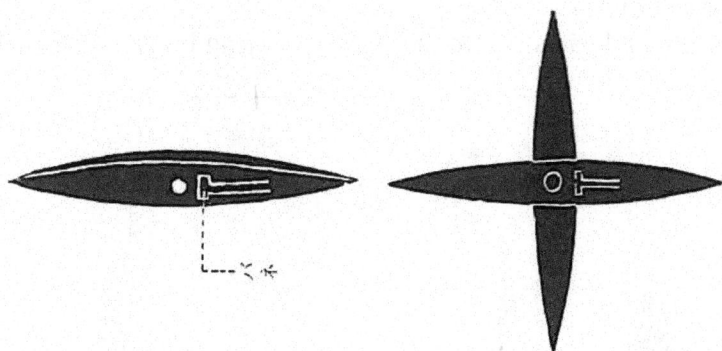

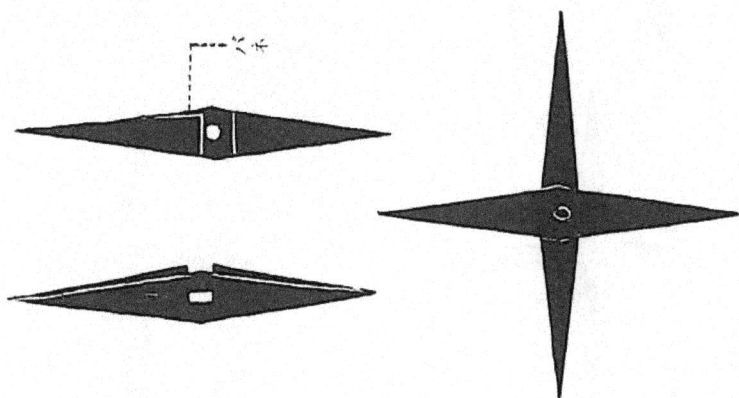

十字形手裏剣

Jujigata Shuriken
Cross Shaped Shuriken

These are Shuriken comprised of two blades that, when opened, form a Juji or cross-shaped Shuriken. There is a spring-lever worked into the design. When closed, it looks like a single blade, but when opened the spring lever locks it into a cross-shaped Shuriken.

The line indicates where the Bane, or spring lever, is located.

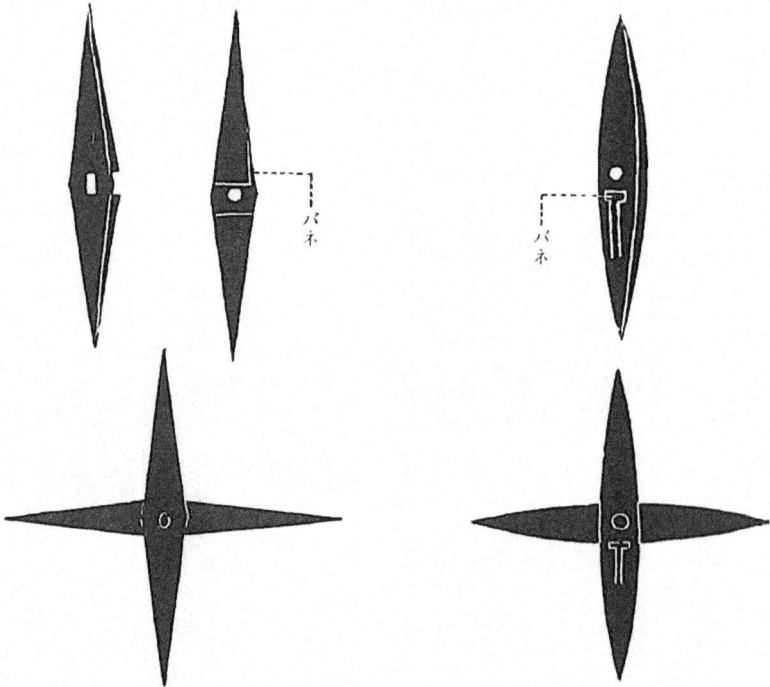

竹内一心流
本流ではこれを
柳枝剣という。

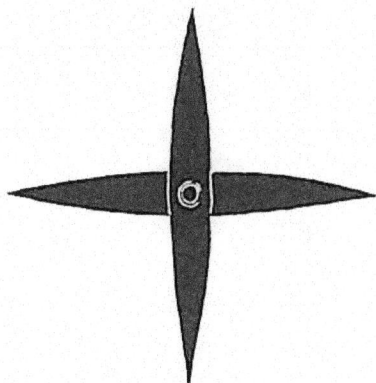

孟淵流
本流ではこれを
宝勝剣という。

153

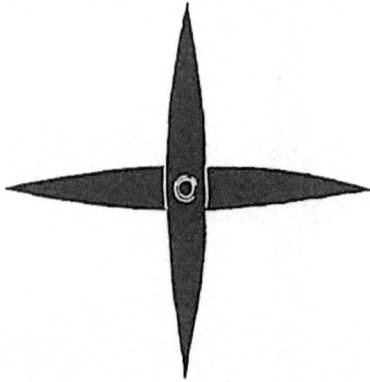

竹内一心流
Takeuchi Isshin School
In the Takeuchi Isshin School this Shuriken is called a Willow Leaf Sword.

孟淵流
Moen School
In the Moen School this Shuriken is called the Treasured Victory Blade

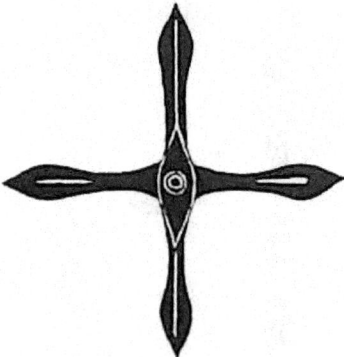

[no information]

手裏剣代用として用いるもの

Other Weapons Used as Shuriken

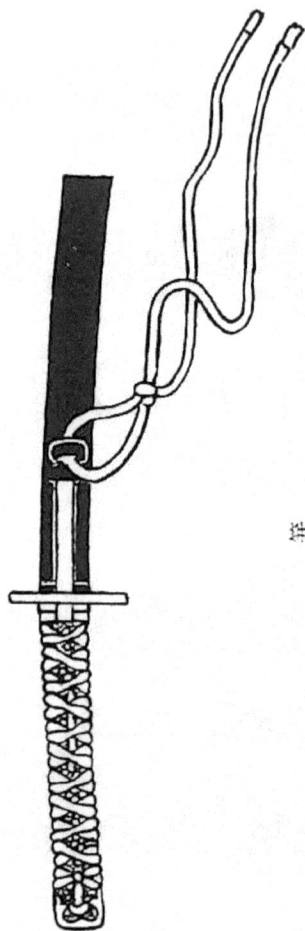

小柄

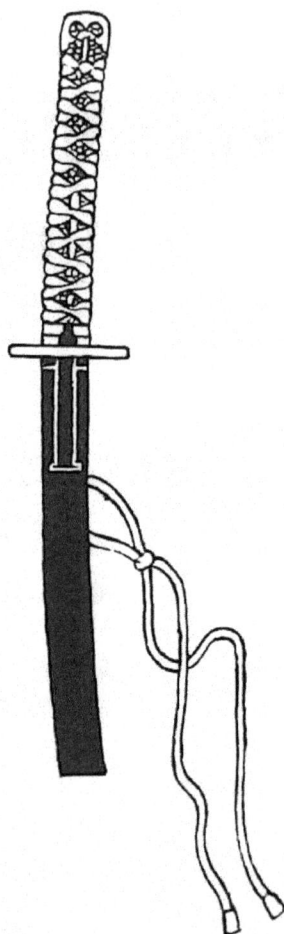

笄

小柄＆笄
Kosuka & Kogai

Translator's Note: The Kozukai, utility knife, and Kogai, hairpin, are stored on either side of the Katana. The Tsuba, hand guard, has 2 holes to accommodate them on either side of the blade.

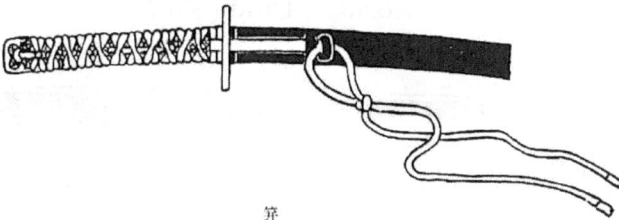

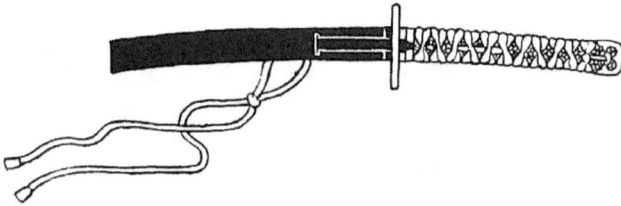

Kozuka

Kogai

Tsuba Hand Guard

Tanto : Knife

短刀

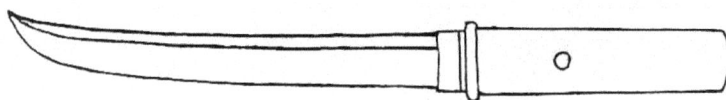

Kozuka : Utility Knife

小柄

Kogai : Hairpin

笄

Kogai : Hairpin

笄

Kogai : Hairpin

笄

Tanken : Knife

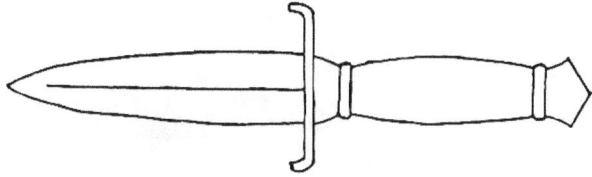

Knife (Written phonetically in Japanese referring to Western style knives.)

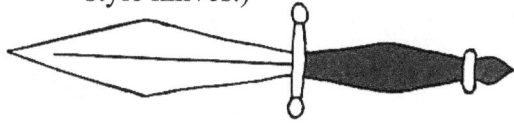

Knife

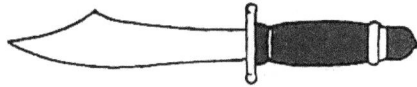

Knife

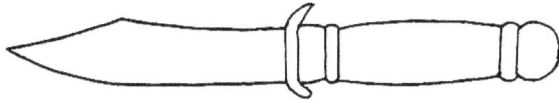

Knife

Knife

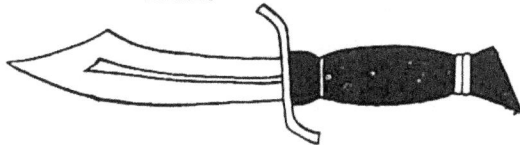

Deba : Japanese Chef's Knife

出刃

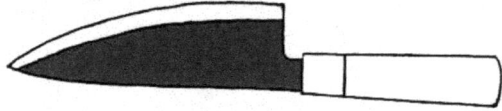

Hibashi : Metal Chopsticks for moving hot charcoal

火箸

Kogai : Hairpin

笄

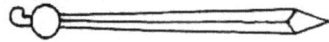

鏢 Hyo : Spear Hairpin

鏢

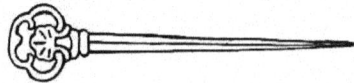

手裏剣の打ち方要領

How to Throw Shuriken: Essential Points

手裏剣を打つには、手裏剣
を、中央中指のところにあ
てて、人差指と無名指で軽
くはさむようにして、拇指
で軽く押さえて持つ。

手裏剣を打ち離す瞬間を指
して離れという。この離れ
は、手裏剣を打つうえに重
大のものであるから、大い
に心得るべきことである。
いずれの指にも平均に軽く
力を入れ（手裏剣をただ押
さえる程度の）、目標に向
かって手裏剣を打ち離すと
き、指は同時に自然と離れ
るように打つ。

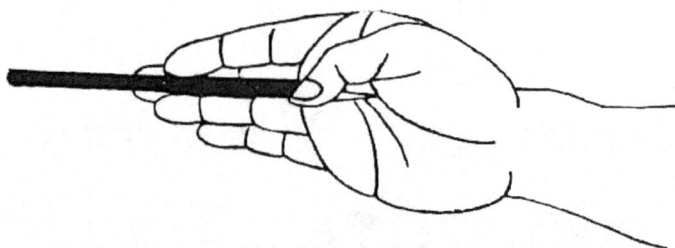

手裏剣の持ち方
How to Hold a Shuriken

When throwing, first align the Shuriken with your middle finger. Lightly support the Shuriken on either side with your index finger and ring finger. Your thumb should press lightly. When throwing, your fingers should break contact at the moment of release. The way the fingers break contact when releasing the Shuriken is the most important part of throwing. Be sure to pay close attention to this. Each finger should apply the same amount of light pressure. This is just enough to hold it in position. When facing your target and throwing all your fingers should release naturally at the same time.

直打法と廻転法

手裏剣の打ち方には、直打法（陽の劔）による打ち方と、廻転法（陰の劔）に

よる打ち方の二種がある。

直打法（陽の劔）とは

剣先を指先の方にして

持ち打つ方法をいい、

point

剣先 …

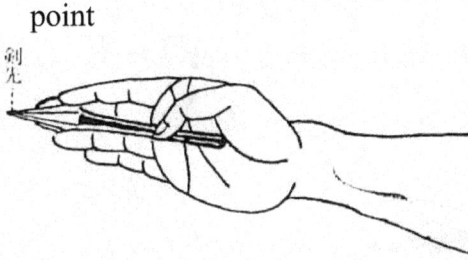

Direct Throw also known as Yang Blade

直 打 法

一 名 陽 の 劔

Choku-uchi Ho 直打法
Kaiten Ho 廻転法
Direct Throw and Revolving Throw

There are two basic categories to throw Shuriken, a Direct Throw, which is Yang Blade, and a Revolving Throw, which is Yin Blade.

Choku-uchi Ho 直打法
Direct Throw
Also known as *Yo no Ken*, Yang Blade

For Direct Throw, Yang Blade, hold the Shuriken facing the ends of the fingers.

廻転法（陰の劔）とは
剣先を掌中の方にして
持ち持つ方法をいう。

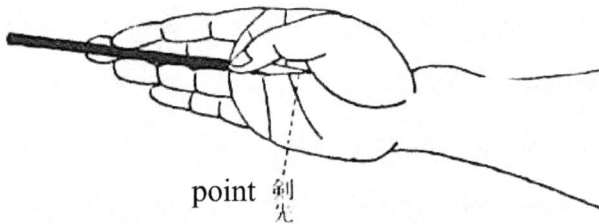

point 剣先

Revolving Throw also known as Yin Blade
廻 転 法
一 名 陰 の 劔

Kaiten Ho 廻転法
Revolving Throw

Also known as *In no Ken*, Yin Blade

For Revolving Throw, Yin Blade, hold the Shuriken with the tip facing your palm.

直打法（陽の劔）は
近距離を打つによく、

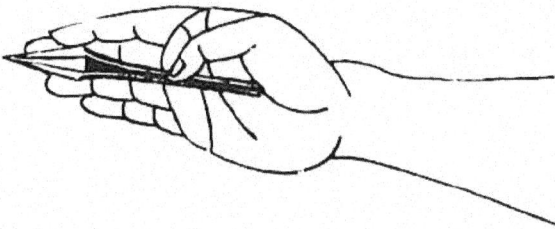

直 打 法

Choku-uchi Ho 直打法
Direct Throw

The Direct Throw, which is Yang Blade, is for when your target is close.

廻転法（陰の劍）は
遠距離を打つによい。

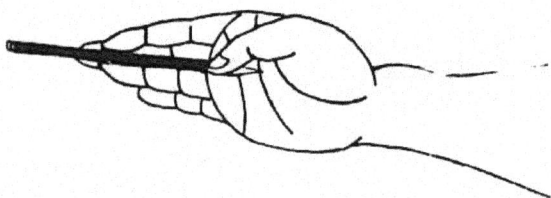

廻 転 法

Kaiten Ho 廻転法
Revolving Throw

The Revolving Throw, which is Yin Blade, is for longer distances.

手裏剣の打ち方

手裏剣の打ち方に三法がある。

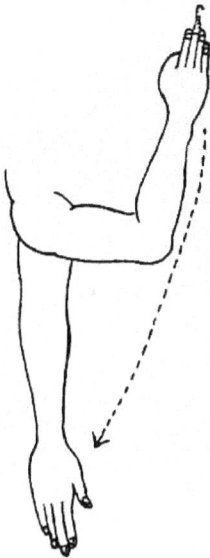

正常打ち　一名　本打ち
陰の打ち

横打ち
中翻打ち
左に打つのを　陽中陰
右に打つのを　陰中陽

逆打ち
陽の打ち

側転打法（陽剣）でも、その打法（陰剣）でも、動作には、正常打（一名本打）と、横打、逆打の三法がある。

正常打は手裏剣本来の打ち方であるが、横打逆打も時に応じて打つ方法で共に練習すべき技である。

167

How to Throw Shuriken

There are three basic throwing techniques.

The two fundamental ways of throwing are the Direct Throw, also known as Yang Blade, and Revolving Throw, also known as Yin Blade. There are also three different throwing motions that are used Normal Throw, Side Throw and Reverse Throw.

Normal throw is the standard way of throwing Shuriken, however depending on the situation you find yourself in, you may need to use Side Throw or Reverse Throw. All three of these techniques should be studied.

	Seijo Uchi 正常打ち Normal Throw Hon Uchi 本打ち Main Throw In no Uchi 陰の打ち Yin Throw
	Gyaku Uchi 逆打ち Reverse Throw Yo no Uchi 陽の打ち Yang Throw
	Yoko Uchi 横打ち Side Throw Chuken Uchi 中剱打ち Center Blade Throw 陽中陰 Throwing with the left is Yang within Yin 陰中陽 Throwing with the right is Yin within Yang.

直打と廻転打による
剣の飛び方の状態

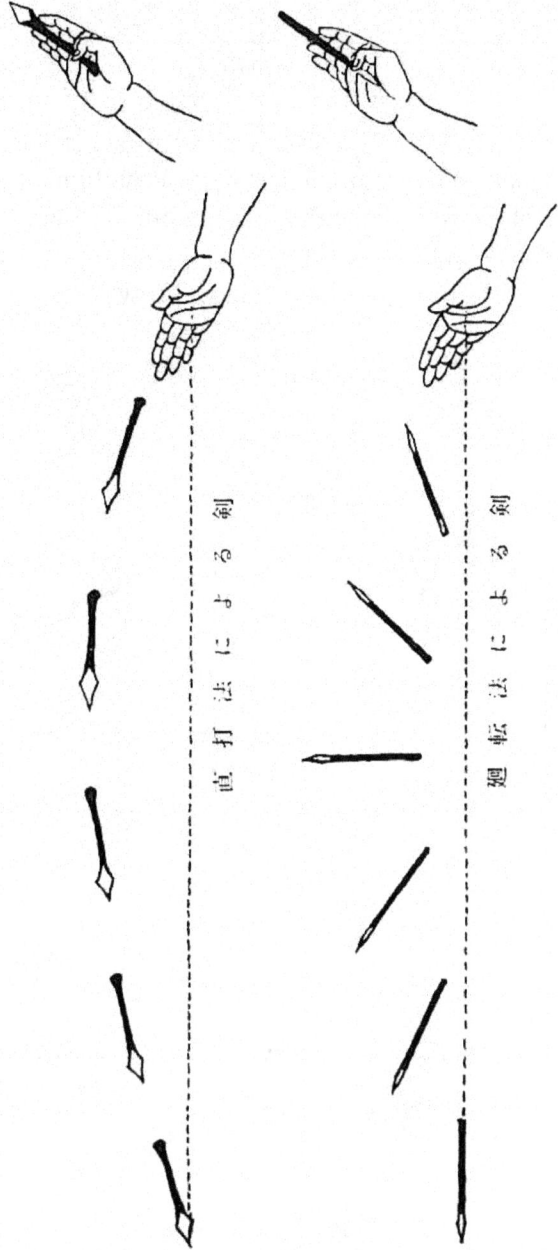

手から離された剣を飛んで行く。
打ちによって、このように飛んで行く。

直打法による剣

廻転法による剣

The path the Shuriken takes to your target differs depending on whether you use Direct Throw or Revolving Throw.
The blade flies through the air in one of these two ways.

Flight Trajectory of Revolving Throw

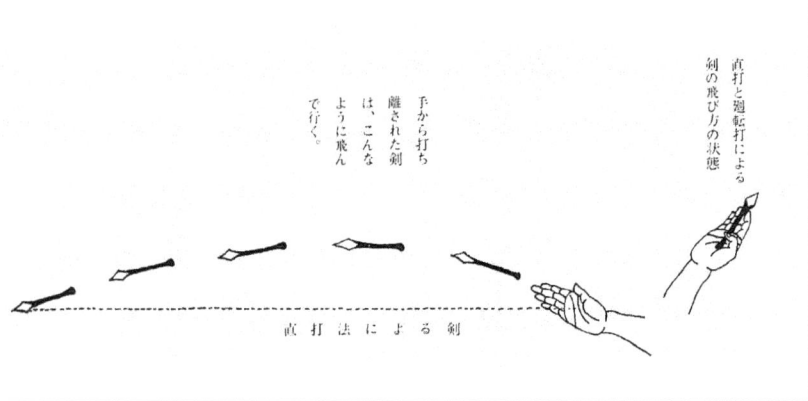

直打と廻転打による剣の飛び方の状態

手から打ち離された剣は、こんなように飛んで行く。

直打法による剣

Flight Trajectory of Direct Throw

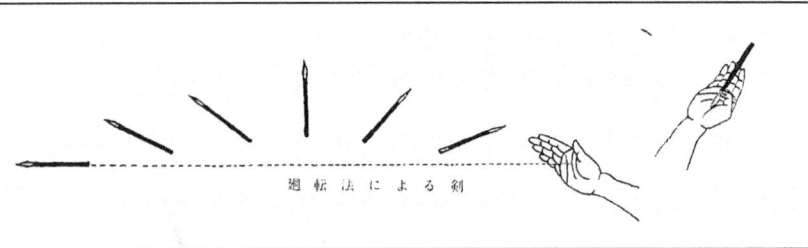

廻転法による剣

手裏剣は、指や手先だけに力を入れて、投げつけるような打方をしてはいけない。剣がグルグルまわって目標に正しく刺さらない。

腕の力を、手掌の外側、小指丘部のところにあつめ、手刀で斬り下ろすような気持で、突き制すように打つ。

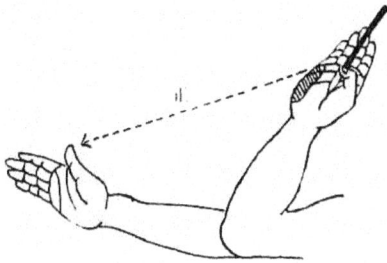

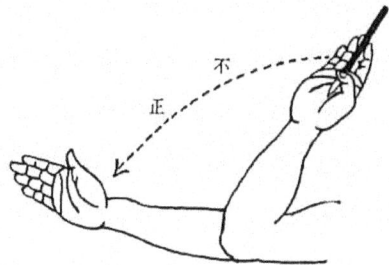

手裏剣打ち方の正と不正
The Right and Wrong Ways to Throw a Shuriken

正 *Sei* Correct	不正 *Fusei* Incorrect
Focus your arm power in the outside edge of your palm at the knuckle of your little finger. Swing your hand down like a Shuto, Knife Hand, as if you are trying to slice something with a sword.	Focusing all your power in the tips of your fingers as you throw is the wrong way. This will cause the blade to spin around and you won't be able to strike your target accurately.

打ち方の要領＝的のねらい方

手裏剣を打つときは、手は目標に向かってまっすぐに差しだして打つがよい。この場合は打たれた手裏剣に、廻転打ちのせ剣先の方向がくるくるやその中心に目標の中心と合致するよう要領で打つ。

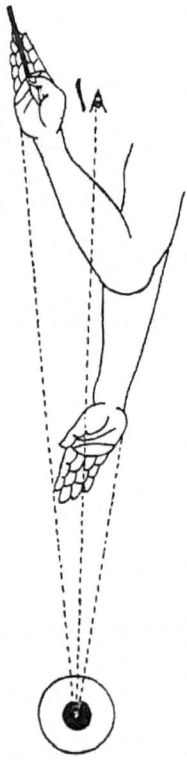

本打ち

手裏剣の打ち方要領

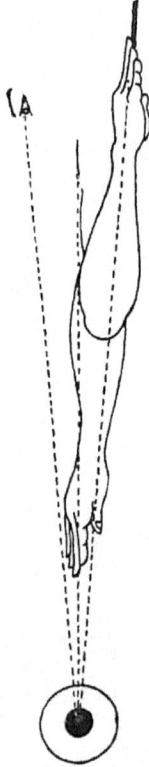

横打ち

ヘ

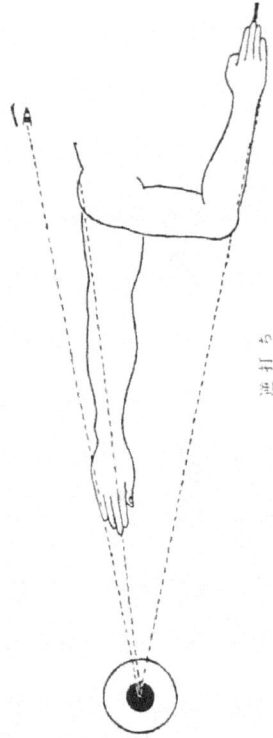

逆打ち

Essentials of Throwing: How to Aim at the Target

When throwing, your hand should extend straight at the target. Focus on lining up the end of the Shuriken in the palm of your hand with your target. If you are doing Direct Throw, then your focus is on the tail end of the Shuriken in your palm. If you are doing Revolving Throw, then your focus is on the point of the Shuriken in your palm. Either way you should ensure that the Shuriken in the center of your palm point lines up with the center of your target.

Hon Uchi Main Throw

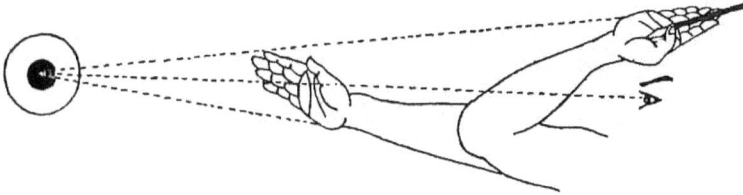

Yoko Uchi Side Throw

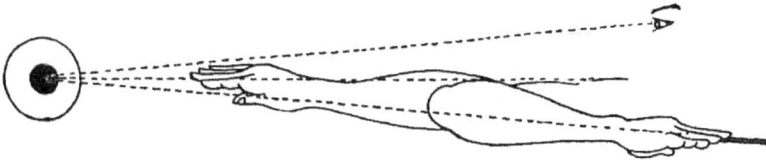

Gyaku Uchi : Reverse Throw

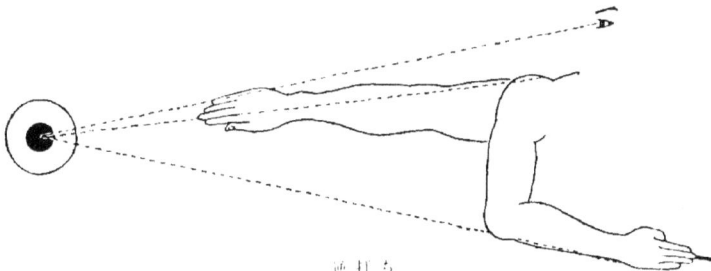

手裏剣の釣合を知ることは、手裏剣を打つうえに重大の結果をもたらすものであるから、図のような方法でよく釣合を調べ、その手裏剣がいずれにも傾むかず、水平になるところが、その手裏剣の中央に当たることを知っておく必要がある。

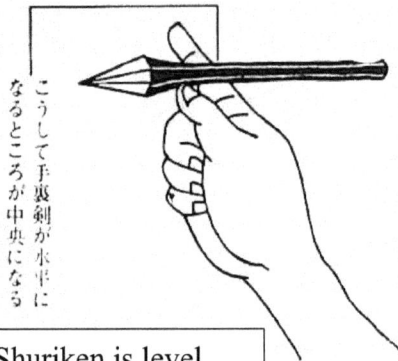

こうして手裏剣が水平になるところが中央になる

Since the Shuriken is level, that means your finger is on the balance point.

手裏剣の釣合を知る方法
How to Check the Balance of a Shuriken

It is important to determine the balance point of your Shuriken. It is very important to find the center of balance of a Shuriken before you throw it. As the illustration shows, position the Shuriken on your finger until it doesn't lean either to the left or right. Once it is level you know you have found the balance point of that Shuriken. This is an essential step.

174

遠近による手裏剣の持ち方

手裏剣を打つ場合、標的の遠近によって、手裏剣の持ち方をかえねばならない。

遠くに打つ場合は、直打ならば、剣先をなるべく中に引いて打ち、廻転法ならば、剣尾をなるべく先に出して打つことが要領である。また、近くに打つ場合は、その反対に、直打ならば、剣先を出し、廻転法ならば、剣尾を引き込めて打つ。

	直打法 Choku Uchi Ho Direct Throw
	廻転法 Kaiten Ho Revolving Throw

遠近による手裏剣の持ち方
Changing Your Hold On the Shuriken According to Distance

You have to change the way you throw Shuriken depending on the distance to your target. The rule of thumb if your target is far away and you are going to use Direct Throw, then pull the tip of the Shuriken in as close to your fingertips as possible.

The rule of thumb if your target is far away and you are going to use Revolving Throw, is to extend the end of the Shuriken out past your fingertips as possible.

If your target is close, then do the opposite. For Direct Throw, extend the tip out further beyond the tips of your fingers and for Rotating Throw, pull the end closer to the tips of your fingers.

遠近いずれでも
剣先、剣尾を延
ばしたり、引っ
込めたりせず、
拇指の屈伸によ
って、遠近度合
を調節すること
が秘伝である。
剣が手の内にな
じまぬときは、
拇指の先でちょ
っと一ひねりす
るようにすると
よくなじむもの
である。

	Adjusting thumb back
	Adjusting thumb forward

    There is a Hiden, or secret teaching, regarding how to adjust
the Shuriken according to the distance. Instead of extending the
Shuriken out or pulling it in, move your thumb forward or back to
adjust for the distance. If you are having trouble handling the
Shuriken, practice using your thumb to rotate it in place. This will
help you learn to handle Shuriken better.

打ち方により的への当たり方が異なる

打剣を強くかませようと思うと、刃先をゆるやかに持って、振りかぶり打つ心構えで打てば、刃の強くかむものである。

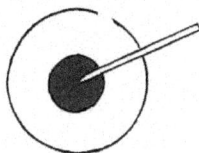

打剣の右よりに立つは、離れに指先のききく故である。

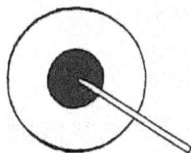

打ち剣左よりに立つは、手のひらにて打つ故で押しつけ離れるためである。

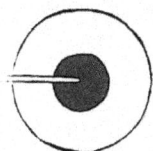

剣の上よりに立つは、離れをおしむためである。手離れをおしむよりに心得打つことが第一である。

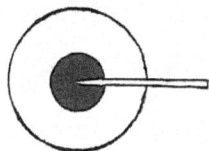

剣がしら下がって立つのは、押しつけのききく故である。

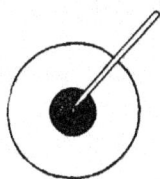

打剣の横半に当たるのは、大指の横にきくるために、つきみがあるためである。

打ち方により的への当たり方が異なる
The Way You Throw Affects the Way the Shuriken Strikes the Target

If you want to throw the Shuriken with power for greater effect, you need to hold the blade loosely in your hand. Then raise it up and throw as if you are trying to hit something and your Shuriken will stick hard in the target.

177

	If the Shuriken you've just thrown strikes in the target leaning to the right, as shown in the illustration, it is because the tips of your fingers are affecting it as you release.
	If the Shuriken you've just thrown strikes in the target leaning to the left, as shown in the illustration, it is because you are throwing with the palm of your hand. You are pushing with your palm as you release.
	If the Shuriken you've just thrown strikes in the target with the point facing down and the end up, as shown in the illustration, it is because you are releasing too late. You need to focus your training on not releasing too late.
	If the Shuriken you've just thrown strikes in the target and hangs down, as shown in the illustration, it is because you are pushing too hard.
	If the Shuriken you've just thrown sticks in the target leaning to the side, it's because your thumb is pushing on the side as you release. You are gripping too hard with your thumb.

短
刀

Tanto

短
刀

Tanto

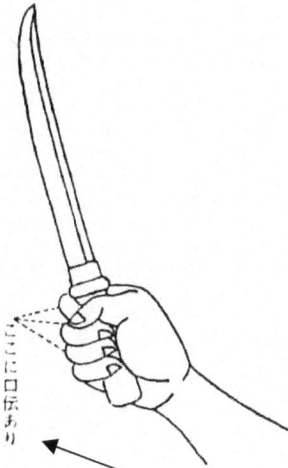

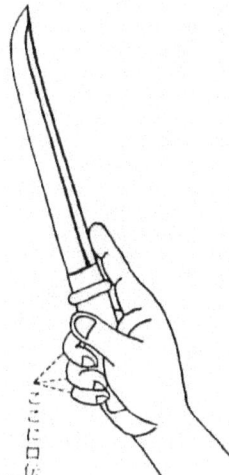

ここに口伝あり

ここに口伝あり

直打法によって短剣を打つ場合の短剣の持ち方

小柄、笄等の類を手裏剣として打つ場合は、廻転打法による打方が多く用いら
れるが、短剣打ちの場合は、直打法による打方が多く用いられ、廻転打法によ
る打方はあまり用いられない。また、片刃の物は、廻転打法も用いられるが、
両刃の物はほとんど廻転打法は用いられず、直打法が用いられる。また、大刀、
小刀の撃ち方は、投げ鎗、打根の撃ち方と同じ要領になるから、別に撃ち方が
ある。

There is a Kuden oral only transmission about how to grip

Ways to Hold the Tanto Knife When doing Direct Throw

Typically when using a Kozuka (utility knife) or a Kogai (hairpin) as a Shuriken, a Rotating Throw is used, however when throwing a Tanken (short sword) the Direct Throw is more common. In fact the Rotating Throw is not usually used with the Tanken. When throwing a single-bladed weapon the Rotating Throw is most common, however for double-bladed weapons the Direct Throw is used in favor of the Rotating Throw. In addition the Tachi (long sword) and Shoto (short sword) have their own throwing method, just like the Toso (throwing spear) and Uchine (throwing dart.)

短
劔

短
刀

Tanken

Tanto

頭指と中指の間に
刀腹をはさみ

拇指で刀腹を押さ
えて持つ

囘転打法によって短剣を打つ場合の短剣の持ち方

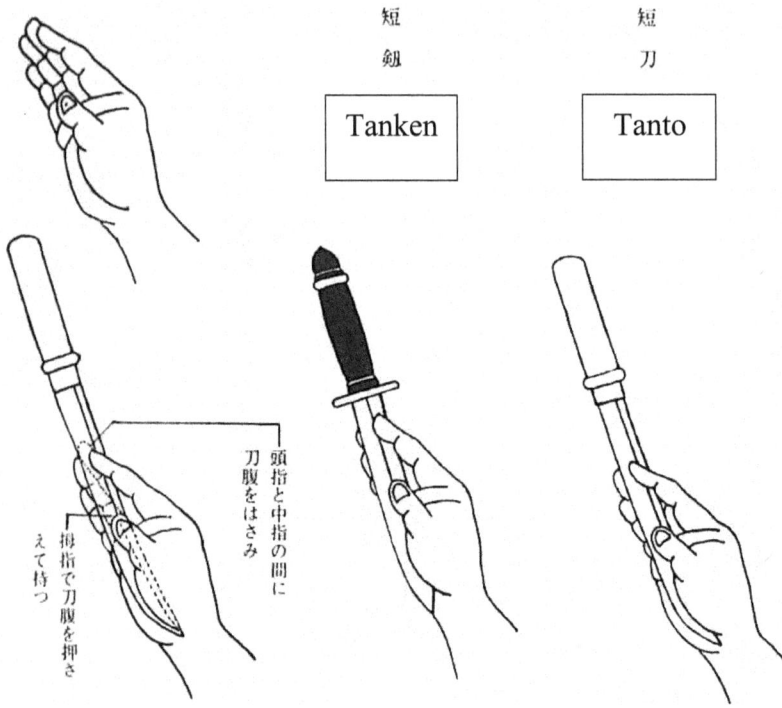

How to Hold the Tanto Knife When doing Rotating Throw

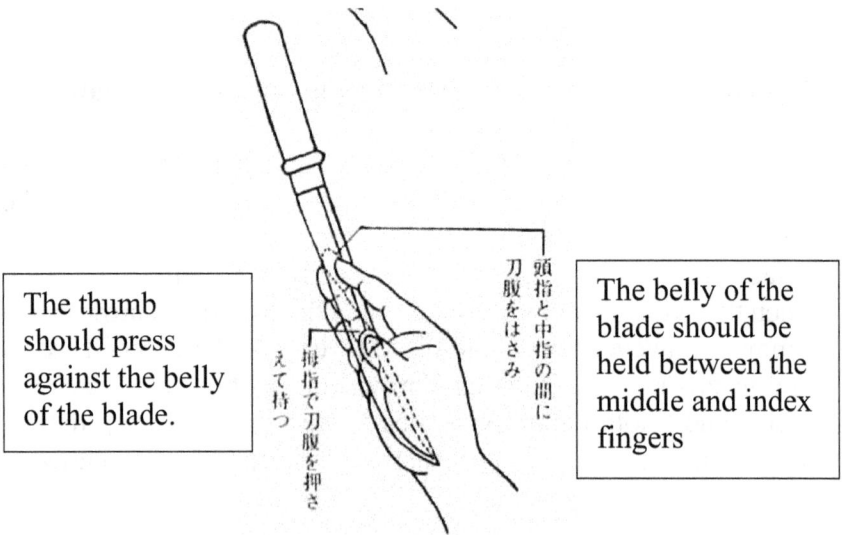

The thumb
should press
against the belly
of the blade.

拇指で刀腹を押さえて持つ

頭指と中指の間に
刀腹をはさみ

The belly of the
blade should be
held between the
middle and index
fingers

View from the back	View from the front

裏より見る　　　　　表より見る

出刃の類を手裏剣として打つときの正しい持ち方

まげた中指の横腹と拇指の腹とで出刃の腹を強く押さえて持つ

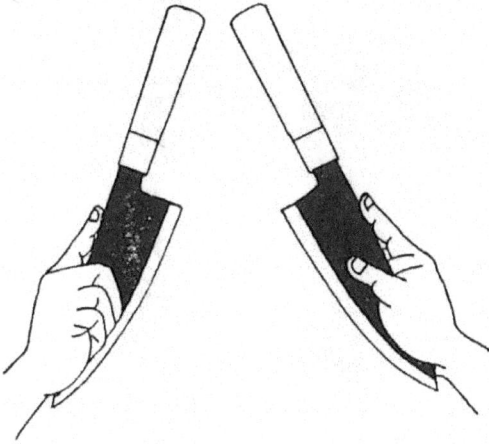

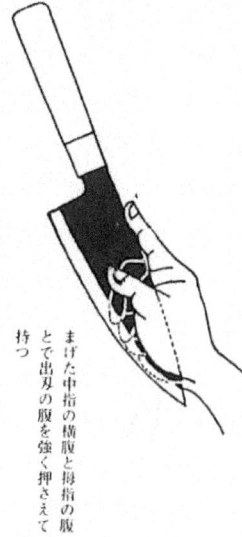

The Correct Way to Hold a *Deba Bocho* Kitchen Knife When Using as a Shuriken

With the blade facing out hold the belly of the blade firmly with your curved middle finger on one side and your thumb on the other.

短
劔

Tanken

両刃の物を手裏剣として打つ場合の持ち方

How to hold a double-bladed weapon when using it as a Shuriken

手裏剣の打ち方要領

四方手裏剣（普通十字手裏剣という）、六方手裏剣、八方手裏剣、十方手裏

剣は、一名これを車剣ともいわれている。

縦横いずれに投げても、グルグルと車のごとく廻転しながら飛んで行き、そ

の一角なり二角なりは必ず突き刺さる様に出来た手裏剣であるからである。

この手裏剣は、打ち方、技術も普通、棒状、針形、釘形、角形、平形、短刀

形、鎗の穂形等のような、習練も要せず、かなりの遠距離でも容易に打ち立て

ることができるものではあるが、目標に的確に突き刺さるようになるには、や

はりそれ相当の練習が必要である。持ち方によっても、飛び方に変化があり、

投げ方が悪いと方向のくるいができる。その点大いに研究工夫の必要がある。

九二

十字手裏剣を打つ場合の持ち方

十字手裏剣の正しくない

持ち方。この持ち方で投

げると、力弱く遠くえも

飛ばず、力を入れすぎる

と、すぐ下方に流れる。

Shiho "Four Direction" Shuriken, also known as Juji "Cross Shaped" Shuriken, along with Roppo "Six Direction," Happo "Eight Direction" and Juppo "Ten Direction" Shuriken are collectively known as Shaken, Wheeled Blades.

It does not matter if you throw this type of Shuriken vertically or horizontally, they will spin as they fly, and one or two points will always stick in your target. In addition, unlike with Shuriken shaped like a stick, needle, nail, triangle, flat, a knife or spearhead you can throw them a long distance and have them stick easily with almost no training. However, being able to strike a target precisely takes a great deal of training. There are different ways to hold and throw these types of Shuriken. If you throw incorrectly then your Shuriken will drift off target. It takes a lot of practice to correct such errors.

183

十字手裏剣を打つ場合の持ち方
十字手裏剣の正しくない
持ち方。この持ち方で投
げると、力弱く遠くえも
飛ばず、力を入れすぎる
と、すぐ下方に流れる。

How to hold a Juji Shuriken

This is an incorrect way to hold a Juji Shuriken. If you do throw this way it will not travel far and strike weakly. On the other hand if you put a lot of power in it, the Shuriken's trajectory will drop down.

<table>
<tr><td>

遠近ともに、的中率が多い正しい持ち方。

</td><td>

遠距離によい正しい持ち方である。

</td><td>

同じく正しくない持ち方で、的中率少なく、下方に外れ易い。

</td></tr>
<tr><td>

This throwing style is accurate for targets both near and far. This way of throwing has high accuracy.

</td><td>

The correct way to hold the Shuriken when throwing a long distance.

</td><td>

This is also an improper way to hold a shuriken. It has a low probability of hitting the target and will tend to deflect off the bottom.

</td></tr>
</table>

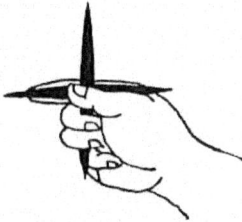

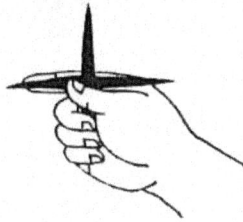

六方手裏剣を打つ場合の
正しい持ち方。

近距離に適する正しい持
ち方。

遠近ともによい正しい持
ち方。

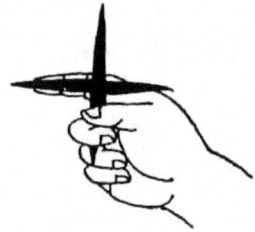

| This is the correct way of holding as six-pointed Shuriken before throwing. | This is a correct way of holding a Shuriken for short distances. | This a correct way of holding a Shuriken suitable for both long and short distances. |

遠距離に適する正しい持ち方。

直距離に適する正しい持ち方。

八方手裏剣の正しくない持ち方、この持ち方は、刀の入れようで、上下いずれかに外れる。

This is the correct way to hold the Shuriken for a long-distance throw.

This is the correct way to hold the Shuriken for a short-distance throw.

This is how not to hold a Happo, Eight-Pointed Shuriken. This way of holding will mean the throwing motion will cause it to either go over or under your target.

車剣（十字、六方、八方等の）手裏剣は、とかくその持ち方と打ち方（手離れ）によって、左え左えとその方向がそれ易くなるものであるから、その点大いに心して、まっすぐ打つように心掛けるべきである。手もとのちょっとしたくるいも、速くなればなるほど大きなくるいとなるものである。

それから手裏剣はいかなる場合でも、持っている手裏剣全部を打ち尽くさず、一つは必ず手に残すのが心得の一つである。それはいざという場合に役立てるためである。

不正

正

十字手裏剣は下図のように握って、第四図のように格闘に備える。それを左図のように握ってそれを投げる。

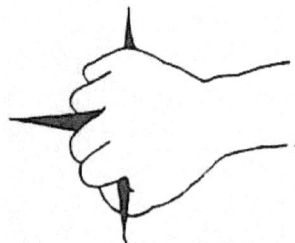

188

When throwing Shaken, Wheeled Blades, which refers to Cross-Shaped, Six-Pointed, Eight-Pointed and other such Shuriken, the way you hold them is important. When releasing the Shuriken they tend to drift either to the left or the right. You have to train diligently in order to correct this tendency. Even a slight imbalance when throwing will be exacerbated over a long distance.

Finally, no matter what situation you are in, never throw all your Shuriken. You should always remember to keep one in reserve. It will come in handy in a difficult situation.

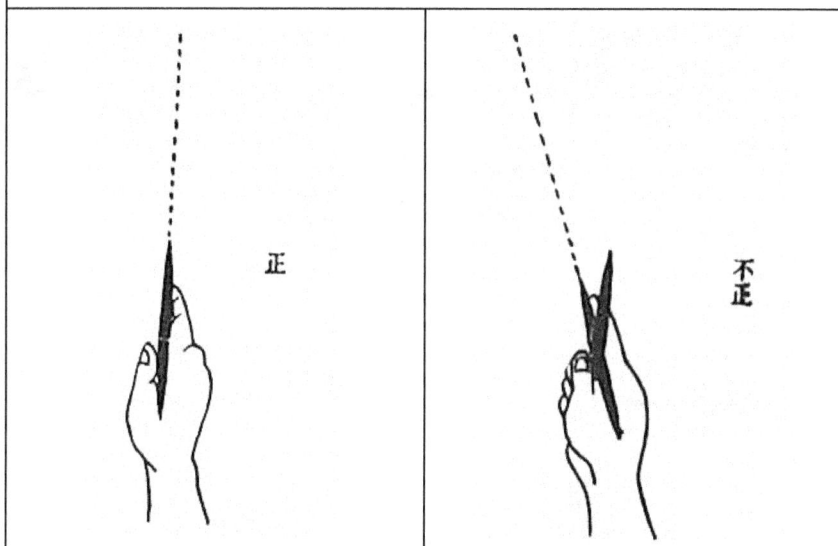

正	不正
Correct	Incorrect

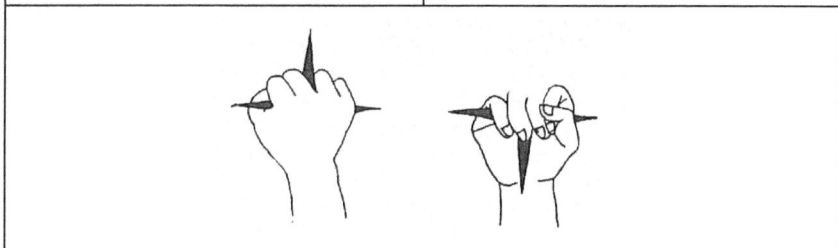

Holding a Cross-Shaped Shuriken as shown in the illustration gives you a weapon in hand-to-hand combat.

手裏剣打ち方練習法
立打ち

How to Throw Shuriken: Standing Throw

標的の造り方 （一）
How to Make a Shuriken Target #1

Width 90 centimeters
Height 174 centimeters
Width 6 cm

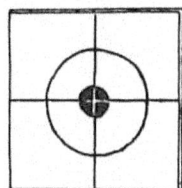

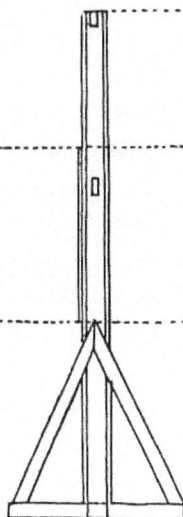

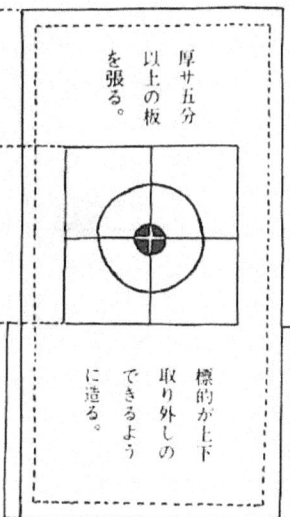

Target	Stand	Door Panel
標　的	板戸立合	板　戸

This target is comprised of three parts: a door panel, a stand and the target.

The target should be a square board 2 Shaku 60 centimeters on a side with a thickness of at least 5 Bun, 1.5 centimeters. Draw a circle in the center with a diameter of 1 Shaku, 90 centimeters. In the very center draw another circle with a diameter of 3 Sun, 9 centimeters and color it in black.

The door panel has a width of 3 Shaku 90 centimeters and a height of 5 Shaku 8 Sun, 174 centimeters. It should be at least 5 Bun, 1.5 centimeters thick. Position the target so that any Shuriken that go high or low will only strike the wooden panel. Make a border of strips of wood 2 Sun, 6 centimeters thick.

標的の造り方 （二）
How to Make a Shuriken Target #2

板のかわりに
古畳を用いる
方が手裏剣が
いたまないで
よい。
寸法その他は
図を見て工夫
すべきである。

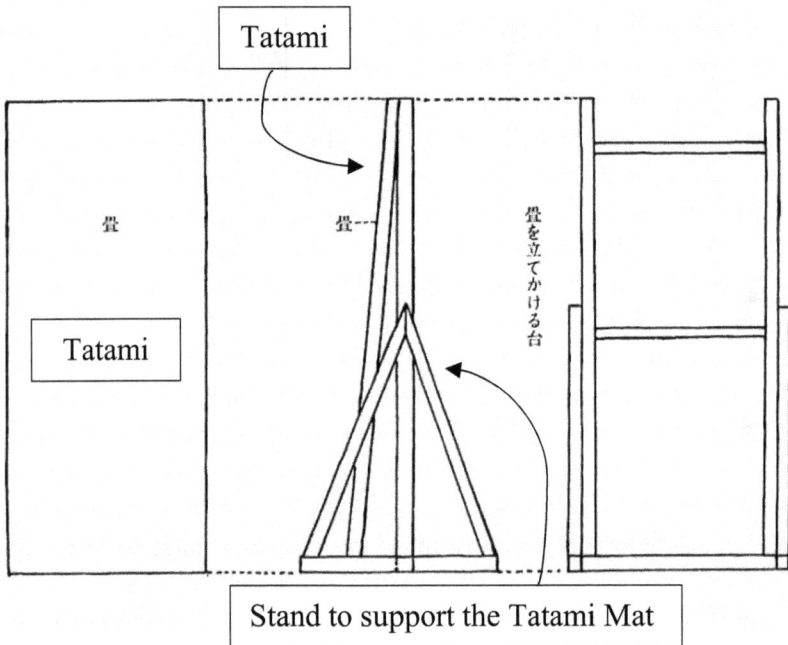

Tatami

畳

Tatami

畳

畳を立てかける台

Stand to support the Tatami Mat

You can also use an old Tatami mat instead of a wooden board. This will prevent your Shuriken from getting damaged. The illustration below gives general instructions, but you may need to make some adjustments.

手裏剣の打ち方練習法（一）立打ち
How to Practice Throwing Shuriken: Method 1 Standing Throw

The illustration shows the best points to aim at in order to topple your enemy.

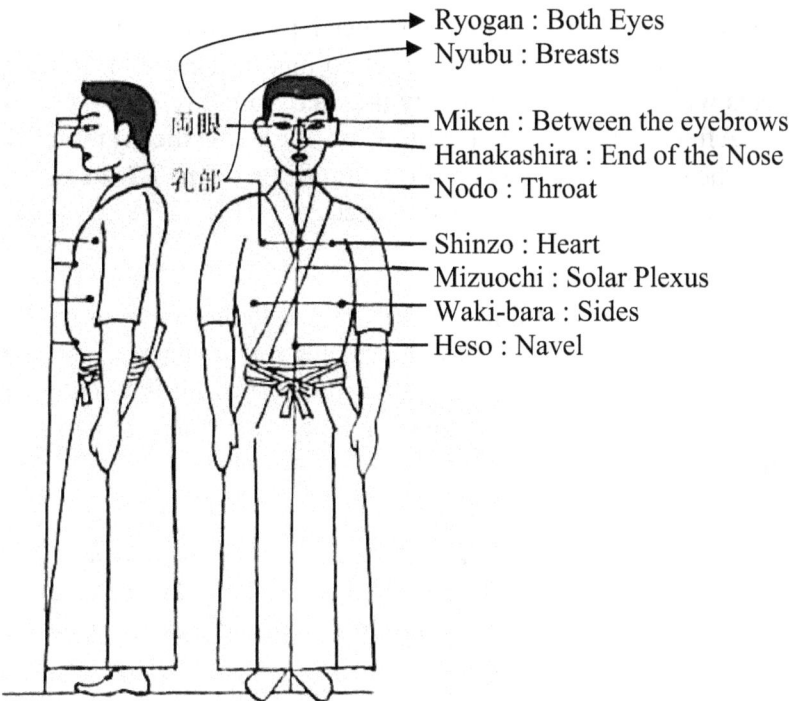

Ryogan : Both Eyes
Nyubu : Breasts

Miken : Between the eyebrows
Hanakashira : End of the Nose
Nodo : Throat

Shinzo : Heart
Mizuochi : Solar Plexus
Waki-bara : Sides
Heso : Navel

手裏剣の打ち方練習法（一）立打ち
How to Practice Throwing Shuriken: Method 1 Standing Throw
How to position yourself in front of the target

標的位置の定め方
練習に先だって、まず標的の位置の定め方であるが、標的はその中心部と、自己の臍とが合致するところを正位置と定め、初めは三間くらいの間隔のところから練習に入るがよい、そしてそれが正確に打てるようになったら、漸次四間、五間、六間と、その間隔を延ばして練習する。

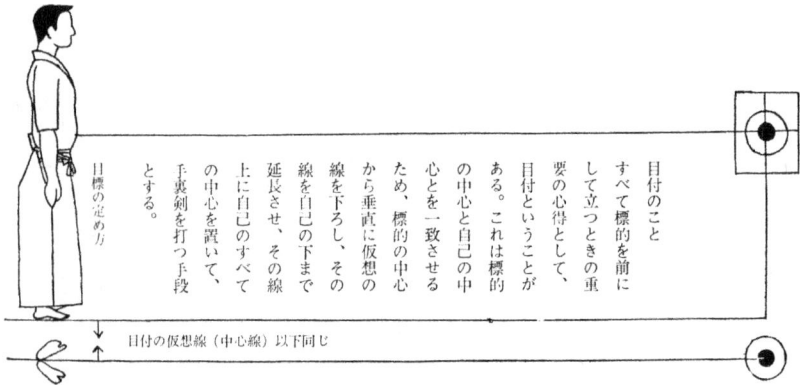

目標の定め方
目付のこと
すべて標的を前にして立つときの重要の心得として、目付ということがある。これは標的の中心と自己の中心とを一致させるため、標的の中心から垂直に仮想の線を下ろし、その線を自己の下まで延長させ、その線上に自己のすべての中心を置いて、手裏剣を打つ手段とする。

目付の仮想線（中心線）以下同じ

Before you begin training you need to position your target properly. You should hang the target so that a line drawn from it intersects with your navel. Initially practice from 3 Ken 5.4 meters away. After you achieve a degree of accuracy then gradually increase the distance, so you are throwing from 4 Ken 7.2 meters, 5 Ken 9 meters and finally 6 Ken 10.8 meters.

How to Aim at Your Target
The most important thing to do when standing in front of the target is aiming properly. In order to ensure you are centered on the target, imagine a line descending straight down from the center of the target to the ground. Then imagine that line extending until it is directly below you.

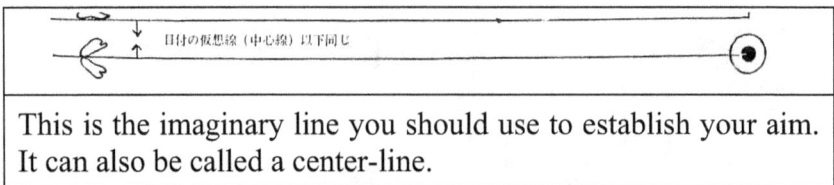

目付の仮想線（中心線）以下同じ

This is the imaginary line you should use to establish your aim. It can also be called a center-line.

手裏剣の打ち方練習法（一）立打ち
How to Practice Throwing Shuriken: Method 1 Standing Throw
練習法第一動作　卍字型
Manji Gata : How to throw : Step 1

Translator's Note: Manji Gata seems to be a basic throwing method that is common in most schools of Shuriken.

Front View Side View

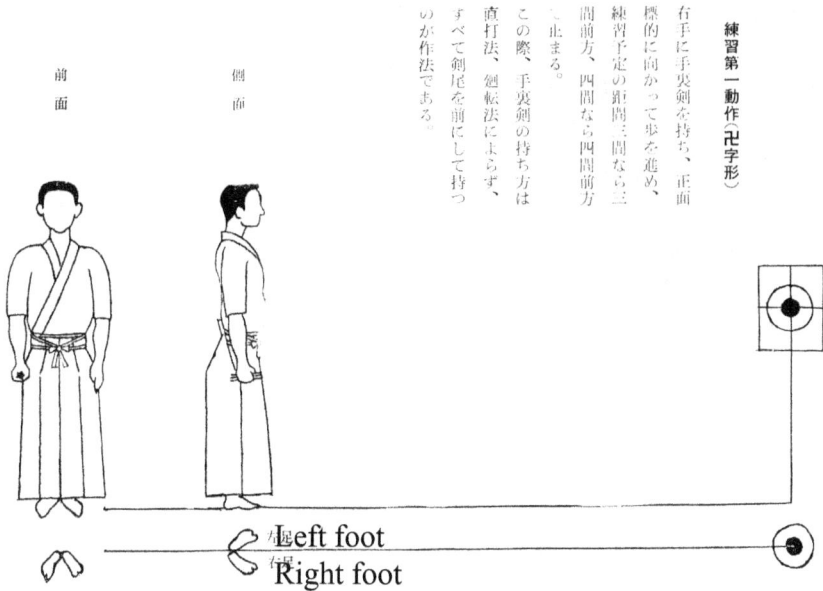

前面　　　　側面

右手に手裏剣を持ち、正面
標的に向かって歩を進め、
練習予定の距離三間なら三
間前方、四間なら四間前方
に止まる。

この際、手裏剣の持ち方は
直打法、廻転法によらず、
すべて剣尾を前にして持つ
のが作法である。

練習第一動作（卍字形）

Left foot
Right foot

Hold the Shuriken in your right hand. Facing the target, walk towards it until you have reached the predetermined throwing distance and stop. If this is 3 Ken, then stop 3 Ken in front of the target, if it is 4 Ken then stop 4 Ken in front of the target.

No matter which way you are going to throw, Direct Throw or Rotating Throw, the etiquette is to hold the Shuriken with the tail-end facing forward.

手裏剣の打ち方練習法（一）立打ち
How to Practice Throwing Shuriken: Method 1 Standing Throw
Manji Gata : How to throw: Step 2

Front   Side

前面

側面

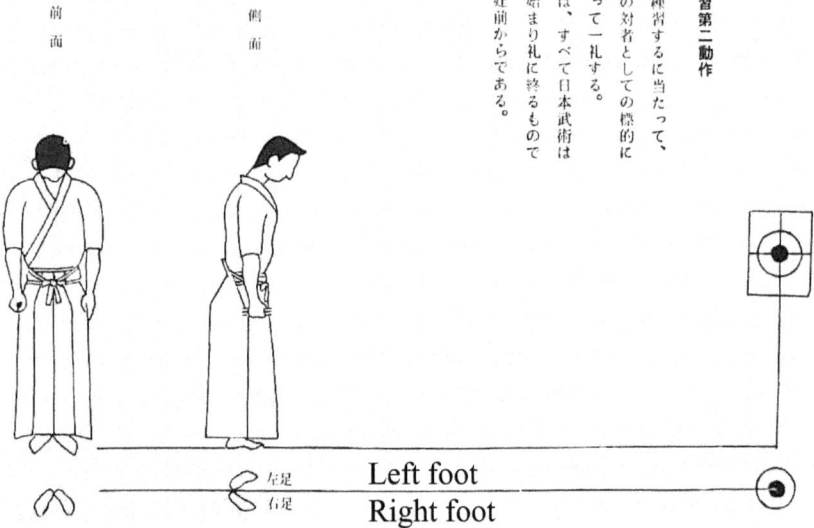

練習第二動作

次に練習するに当たって、仮想の対者としての標的に向かって一礼する。これは、すべて日本武術は礼に始まり礼に終るものである建前からである。

左足
右足

Left foot
Right foot

The next step is to face the target, which represents you opponent, and do a Rei, bow of respect. All Japanese martial arts begin and end with a Rei.

手裏剣の打ち方練習法 （一） 立打ち

How to Practice Throwing Shuriken: Method 1 Standing Throw
Manji Gata : How to throw: Step 3

Front   Side

前面

側面

練習第三動作

的の中心と自己の中心と
を一致させた目附けの一線
を中心に、爪先を閉じる。

左足Left foot

右足Right foot

Ensure that the center of the target lines up with the center of your body to ensure you are aiming properly, then bring your toes of your feet together.

手裏剣の打ち方練習法（一）立打ち
How to Practice Throwing Shuriken: Method 1 Standing Throw
Manji Gata : How to throw: Step 4

Front   Side

前面   側面

練習第四動作

左足拇指で線上を踏み、右
足を線に添って後え引いて
同じく拇指で線を踏み、構
える。

Right Foot   Left Foot

右足   左足

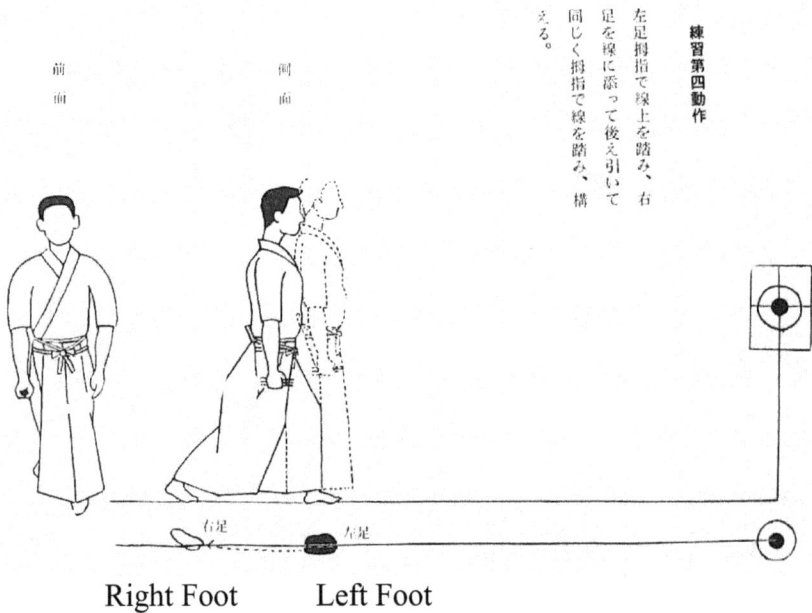

Step forward with your left foot so your big toe is directly on the imaginary line extending from the center of your target. Keep the big toe of your right foot on the same line as you pull your heel back. This is the body positioning after one step.

198

手裏剣の打ち方練習法（一）立打ち
How to Practice Throwing Shuriken: Method 1 Standing Throw
Manji Gata : How to throw: Step 5

Front   Side

前面   側面

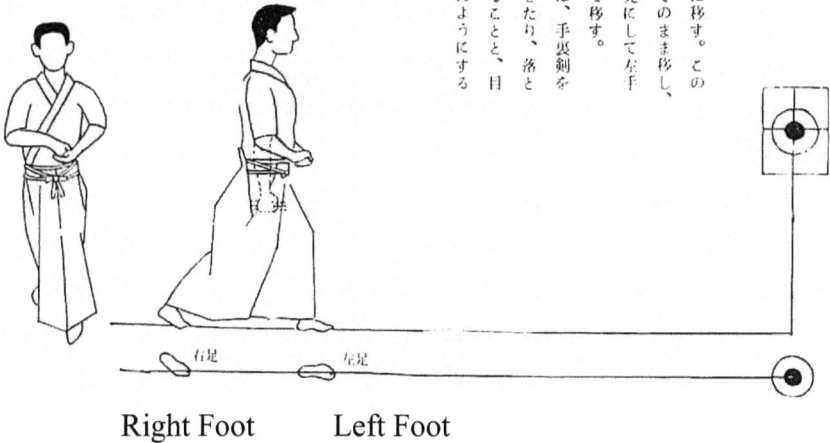

練習第五動作

右手の手裏剣を左の手に移す。この際、廻転法の場合は、そのまま移し、直打の場合は、剣先を先にして左手人差指のところに受けて移す。

このとき注意することは、手裏剣をガチャガチャ触れ合わせたり、落としたりしないようにすることと、目は決して標的から離さぬようにすることである。

Right Foot   Left Foot

Transfer the Shuriken in your right hand to your left hand. If you are going to be throwing Revolving Throw, then transfer the Shuriken in the same way you are holding them in your right hand. If you are going to do Direct Throw, then rotate the points forward and place them on the index finger of your left hand.

手裏剣の打ち方練習法 (一) 立打ち
How to Practice Throwing Shuriken: Method 1 Standing Throw
Manji Gata : How to throw: Step 6

Front　　　　　　Side

練習第六動作

手裏剣を右手から左手に移し
終ったら、右手は静かに下に
たろし、左手はそのまま前腕
に構え、いつでもその手の中
にある手裏剣が右ちに右手で
取り得られるようにして置く。

Right Foot　　　Left Foot

After transferring the Shuriken from your right hand to your left hand, slowly lower your right hand while keeping your left elbow by your side and your left hand extended holding the Shuriken. You should be ready to immediately take a Shuriken from your left hand with your right hand.

手裏剣の打ち方練習法（一）立打ち

How to Practice Throwing Shuriken: Method 1 Standing Throw
Manji Gata : How to throw: Step 7

Front   Side

前面   側面

練習第七動作

次に、左手も手裏剣を持ったまま静かに下におろし、標的と自己との中心を見定めながら、右手で左手の手裏剣をとって持つ。

Right Foot   Left Foot

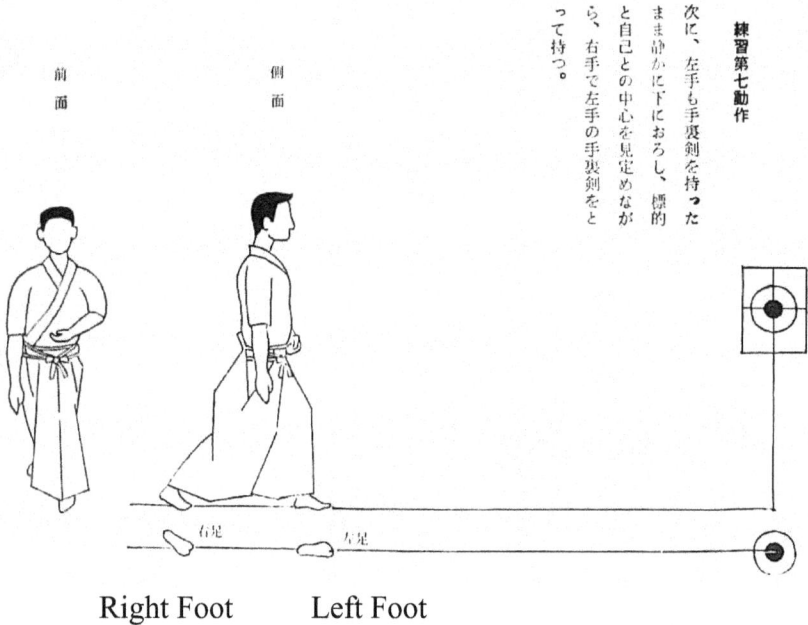

Next, having lowered your right arm silently and holding the Shuriken held in your left hand, focus on how your body is centered on the target. While doing this, take a Shuriken from your left hand with your right hand.

201

手裏剣の打ち方練習法（一）立打ち
How to Practice Throwing Shuriken: Method 1 Standing Throw
Manji Gata : How to throw: Step 8

Front　　　　　Side

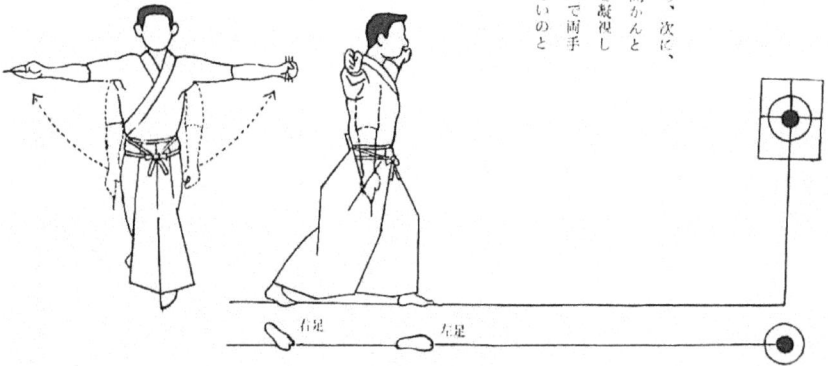

練習第八動作

手裏剣を右手に持ったら、次に、驚かまさにその両翼を開かんとするような体勢で標的を凝視しながら、身造ろい、次いで両手を左右肩と平均するくらいのところまであげる。

Right Foot　　Left Foot

After taking a Shuriken in your right hand stare steadily at your target as you get ready to raise both arms up in a sudden motion. This action of "opening both wings" is to startle your opponent. When you are ready, raise both arms until they are level with the ground.

手裏剣の打ち方練習法（一）立打ち
How to Practice Throwing Shuriken: Method 1 Standing Throw
Manji Gata : How to throw: Step 9

Front   Side

前   倒
面   面

練習第九動作

左右に延ばした手をそのまま左
から右え、身体とともに廻転、
目附けの一線上に一致せしめる。

右足   左足

Right Foot   Left Foot

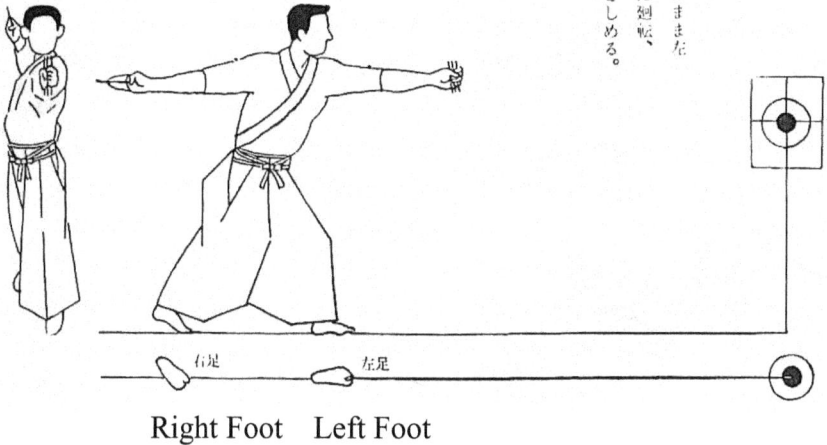

Your arms are extended out to the left and the right. Keeping that body position rotate your torso clockwise until your body is aligned with the imaginary line extending from the center of the target to your navel.

手裏剣の打ち方練習法（一）立打ち
How to Practice Throwing Shuriken: Method 1 Standing Throw
Manji Gata : How to throw: Step 10

Front   Side

前面   側面

練習第十動作

次に、右手の手裏剣をおもむろに右後頭部のやや上方に構え、左手は打たんとする目標に向けて指示するように見当をつける。

右足   右足

Right Foot   Left Foot

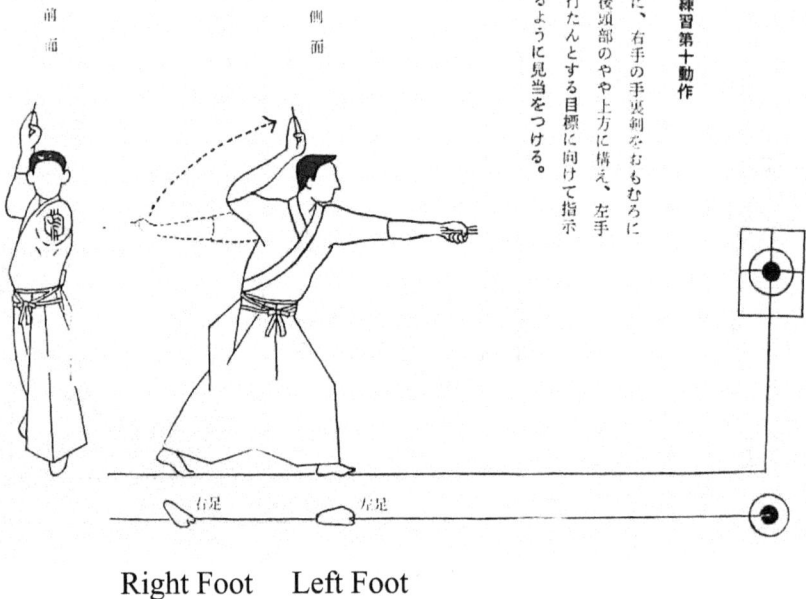

Next raised your right hand slowly until it is behind your head, slightly higher than your head. Keep your left hand pointed at the target you are going to throw towards, as if it is directing the throw.

手裏剣の打ち方練習法 （一） 立打ち

How to Practice Throwing Shuriken: Method 1 Standing Throw
Manji Gata : How to throw: Step 11

Front　　　　Side

前面

側面

練習第十一動作

打つ。
このとき左手は、自然に後ろに
引き下がるようになる。

Right Foot　　Left Foot

右足　　　左足

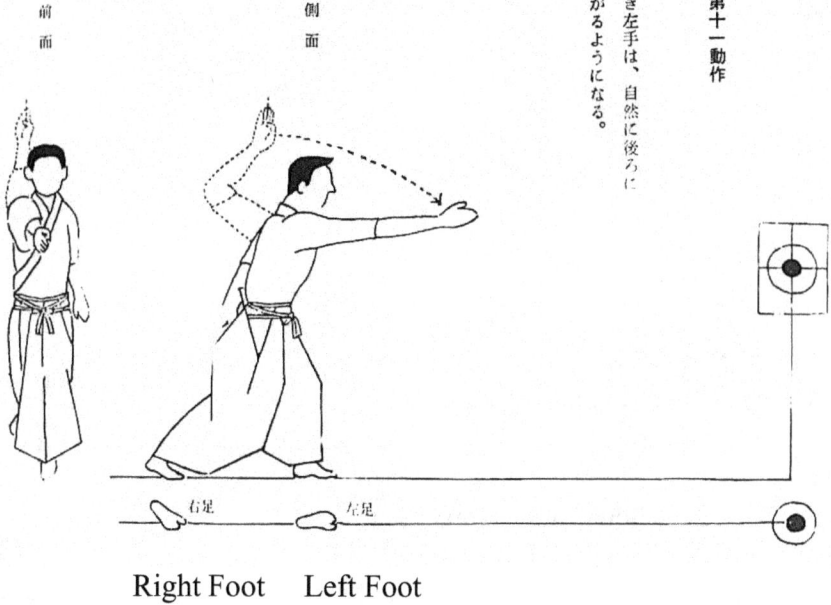

The throw. Your left hand should drop back behind you naturally.

手裏剣の打ち方練習法（一）立打ち
How to Practice Throwing Shuriken: Method 1 Standing Throw
Manji Gata : How to throw: Step 12

Front   Side

前面   側面

練習第十三動作
打ち終ったら、おもむろに
左右の手をもとに戻し、

右足   左足

Right Foot   Left Foot

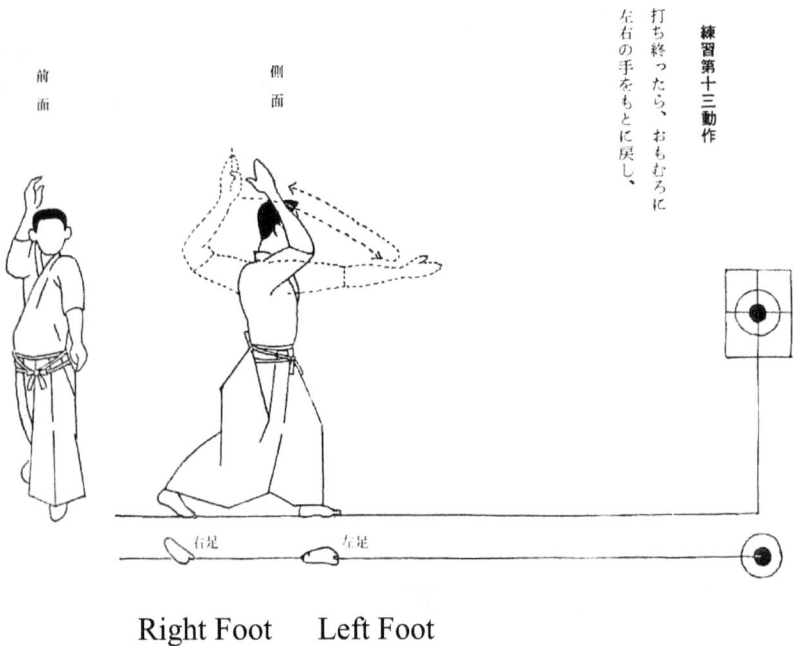

After your Shuriken has hit the target raise your right hand up naturally so that it is slightly above the right side of your forehead.

Translator's Note: This is listed as "Step 13" but clearly comes before the illustration listed as "Step 12." I have corrected the order.

手裏剣の打ち方練習法 （一） 立打ち
How to Practice Throwing Shuriken: Method 1 Standing Throw
Manji Gata : How to throw: Step 13

Front   Side

側面

前面

練習第十二動作

打った手裏剣が的に当たると、その打った右手は自然に右前額上方に上がる。

右足　　　左足

Right Foot   Left Foot

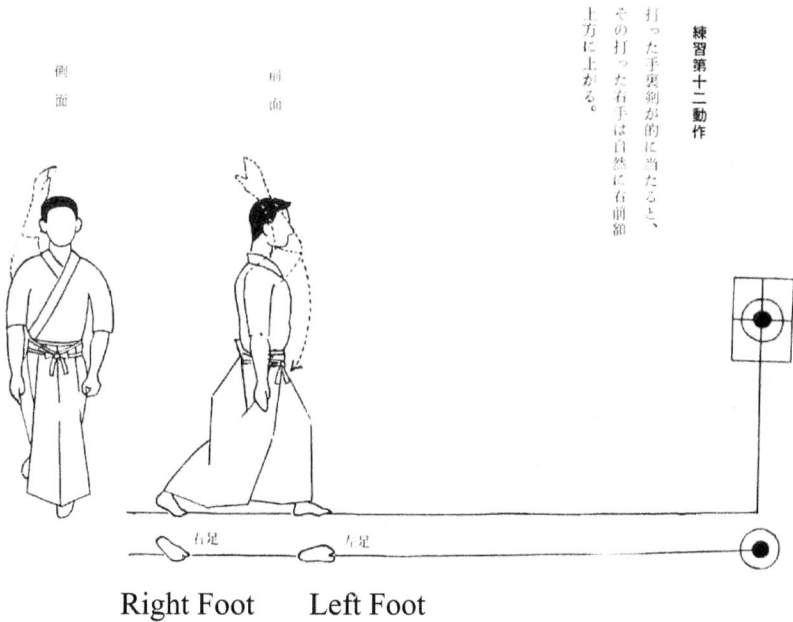

After you have thrown all your Shuriken, slowly return your hands to their starting positions.

手裏剣の打ち方練習法（一）立打ち
How to Practice Throwing Shuriken: Method 1 Standing Throw
Manji Gata : How to throw: Step 14

Front   Side

前面   前面

右足を前方に返して、最初の姿勢となり、

練習第十四動作

左足

右足

Right Foot   Left Foot

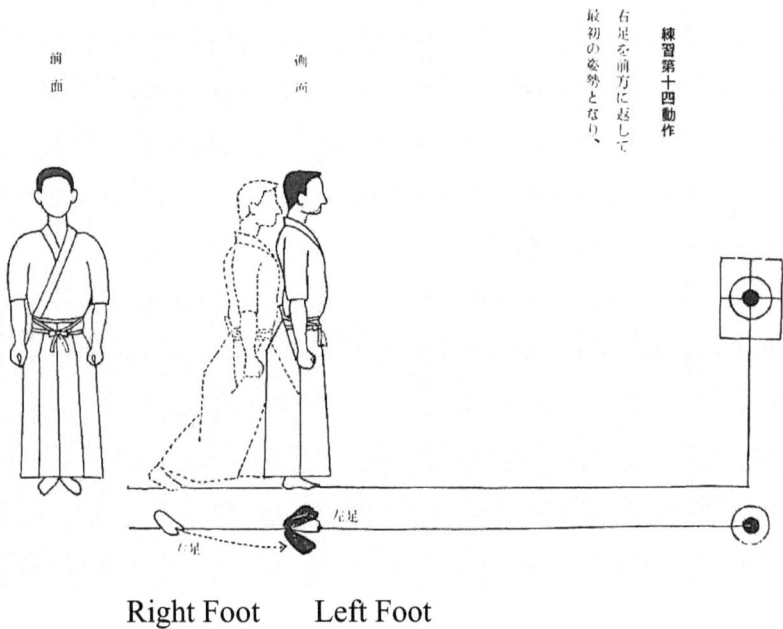

Rotate your right foot forward and move it up beside your left foot. You are now back in the starting stance.

手裏剣の打ち方練習法 (一) 立打ち
How to Practice Throwing Shuriken: Method 1 Standing Throw
Manji Gata : How to throw: Step 15

Front   Side

前面   側面

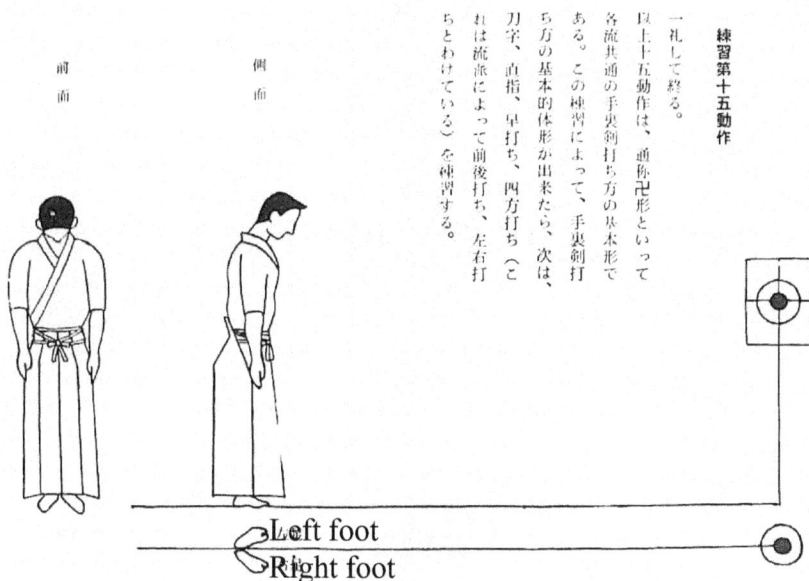

練習第十五動作

一礼して終る。

以上十五動作は、通称卍形といって各流共通の手裏剣打ち方の基本形である。この練習によって、手裏剣打ち方の基本的体形が出来たら、次は、刀字、直指、早打ち、四方打ち(これは流派によって前後打ち、左右打ちともわけている)を練習する。

Left foot
Right foot

Do a Rei, bow of respect, and the technique is over.

This is the end of the fifteen step process known as Manji Gata, which is a standard throwing method across all schools of Shuriken. Once you have trained this basic method of throwing Shuriken you should move on to other styles of throwing such as:

Toji (Sword Shaped)
Jikiyubi (Finger Throw)
Haya Uchi (Rapid throw)
Shiho Uchi (Four Direction Throw)

Four Direction Throw is also known as Front and Back Throw as well as Left Right Throw. The name can vary according to the school.

手裏剣の打ち方練習法 （一） 立打ち
How to Practice Throwing Shuriken: Method 1 Standing Throw
Toji Gata : Katana Style Throw

以上、八動作が刀字形で、上段四動作を一拍子に、下段四動作を一拍子に、都合二拍子に行なうのである。

卍字形は、前掲十五動作の構え（足踏）から打ち終りまでの十二動作を三拍子に行なうのであるが、その中の左右に両手を上げる動作（身造ろい）を略し、直ちに左手を前に右手に剣をとって構える動作（矩）より残心までの八動作を二拍子で行なうのが刀字というのである（重複するが、図示すると次のとおり）

刀字形

The Manji Gata consisted of 15 steps from beginning to end, with the main 12 steps being done with a 1-2-3 rhythm. However, in Katana Style Throw, or Throwing a Shuriken as if you are drawing your sword, the raising the arms to either side element should be eliminated. In Katana Style Throw you will take a Shuriken from your left hand with your right hand and in 8 steps go from throw to Zanshin, with a 1-2 rhythm. Zanshin is maintaining your state of awareness after a technique has finished.

The illustrations show the steps of Katana Style Throw, some of which are repeated from Manji Gata. This ends the eight step throwing process of Togi Gata, Katana Style Throwing.

The top row of illustrations is meant to be done in one movement while the bottom four illustrations represent another movement. These should be done in a 1-2 rhythm.

手裏剣の打ち方練習法（一）立打ち
How to Practice Throwing Shuriken: Method 1 Standing Throw
Jiki Yubi Kata: Finger Throw

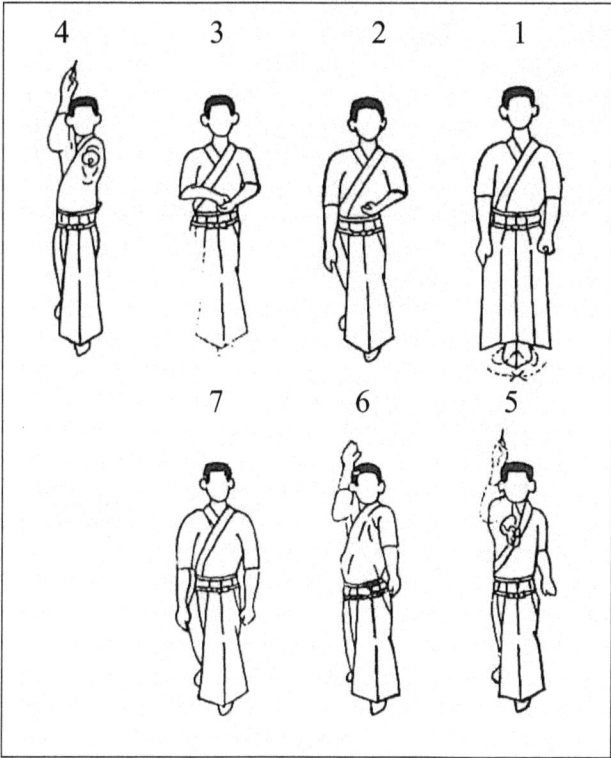

直指形は、刀字形で行なう形をさらに略し、初めから剣を持っていて、一拍子に打つ打ち方で、左図七動作を一拍子に打つのである

直　指　形

The Finger Throw is an abbreviated version of the Katana Throw. You start of holding the Shuriken in your left hand, draw and throw to the count of one. This is a one-beat throw done as shown in the 7 illustrations.

211

手裏剣の打ち方練習法（一）立打ち
How to Practice Throwing Shuriken: Method 1 Standing Throw
Shiho Uchi : Four Way Throw Illustrations 1 - 5

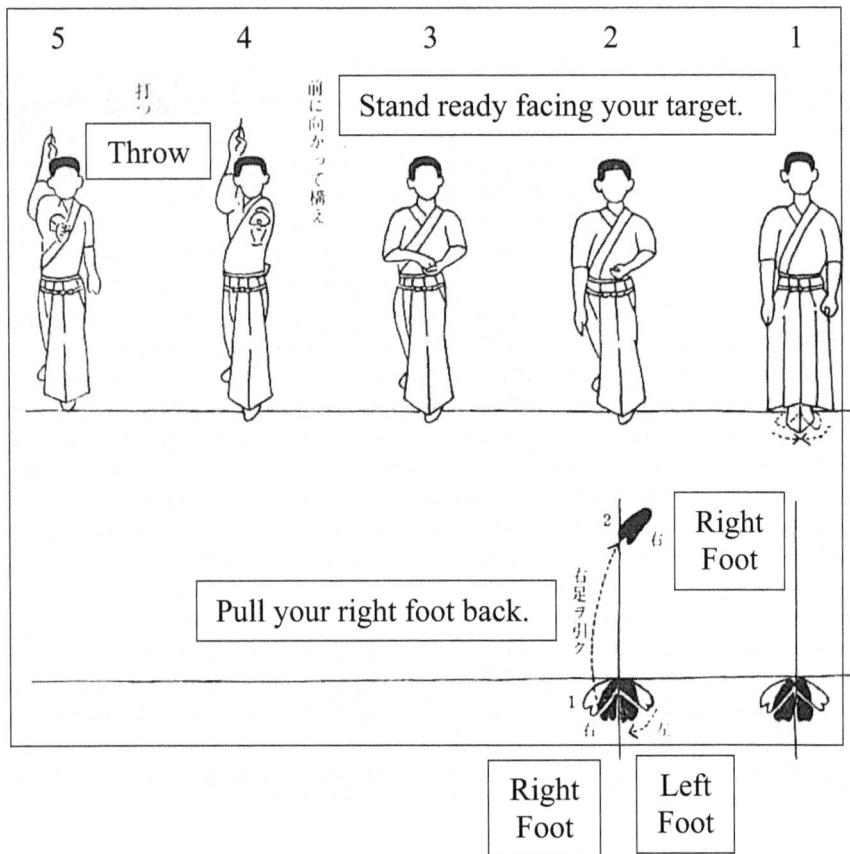

5	4	3	2	1

打つ

前に向かって構え

Stand ready facing your target.

Throw

Right Foot

右足ヲ引ク

Pull your right foot back.

Right Foot

Left Foot

四方打ち

これは、前後左右を打つ練習で、基本体形の運用動作練習である。

This method is for training how to throw in front, behind, to your left and to your right.
This uses the basic way of throwing
This is how to throw forward

手裏剣の打ち方練習法 （一） 立打ち
How to Practice Throwing Shuriken: Method 1 Standing Throw
Shiho Uchi : Four Way Throw Illustration 6 - 9

Throw	Face left and ready,	Throw	Turn around and ready,

9　　　　　8　　　　　7　　　　　6

打つ　　左に向いて構え　　打つ　　後に向いて構え

Rotate your left foot forward.

Rotate your left foot.

Rotate your right foot and pull it back.

In a twisting motion turn your right foot around and pull it back.

手裏剣の打ち方練習法（一）立打ち

How to Practice Throwing Shuriken: Method 1 Standing Throw

立打ちによる本打ち、横打ち、逆打ちの練習

How to Train :

Hon Uchi Main Throw

Yoko Uchi Side Throw

Gyaku Uchi Reverse Throw

逆打ち　　　　　　　　　横打ち　　　　　　　　　本打ち

| Rotate your left foot. | Rotate your left foot. | Rotate your left foot. |

 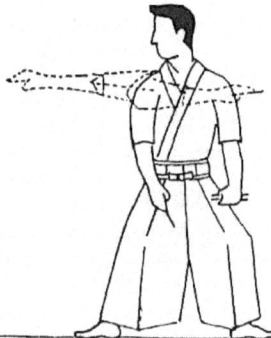 

Reverse Throw　　　　　Side Throw　　　　　Main Throw

手裏剣の打ち方練習法（一）立打ち
How to Practice Throwing Shuriken: Method 1 Standing Throw
歩行短剣逆打ち
How to Throw a Tanken, Short Sword, With a Reverse Throw While Walking (Right Side)

This is how to throw when you are walking, and an enemy appears on your right side. See the detail illustration below.

手裏剣の打ち方練習法 （一） 立打ち

How to Practice Throwing Shuriken: Method 1 Standing Throw

歩行短剣逆打ち

How to Throw a Tanken, Short Sword, With a Reverse Throw
While Walking (Left Side)

This is how to throw when you are walking and an enemy appears on your left side. See the detail illustration below.

Translator's Note:
The martial artist and researcher Nawa Yumio 名和弓雄 (1913-2006)included Fujita Seiko's Shuriken target designs in his 1974 book 図解隠し武器百科 Illustrated Encyclopedia of Concealed Weapons. He does not specify which schools used the following targets.

Target 1: *Domado* Moving Target

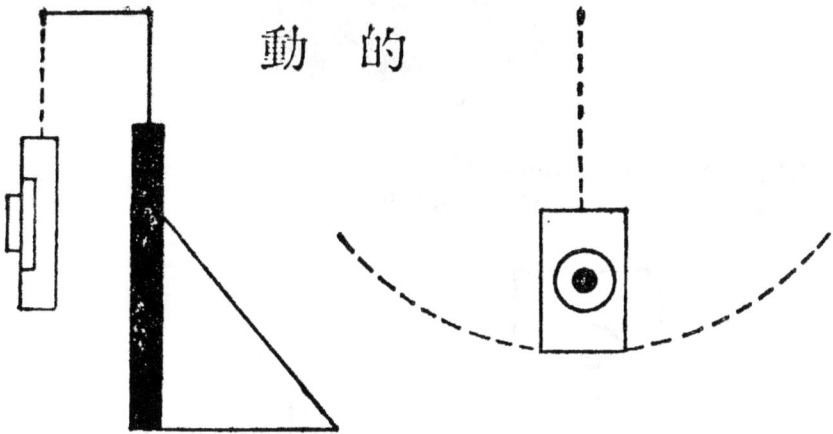

動　的

Target 2 :*Kinmado* Close Target
The large circle has a diameter of 5 Sun, 15 centimeters.
The smaller circle has a diameter of 2 Sun, 6 centimeters.

近 的
大5寸
小2寸

Target 3: *Hashiri nagara Utsu*
Targets to throw at whilst running

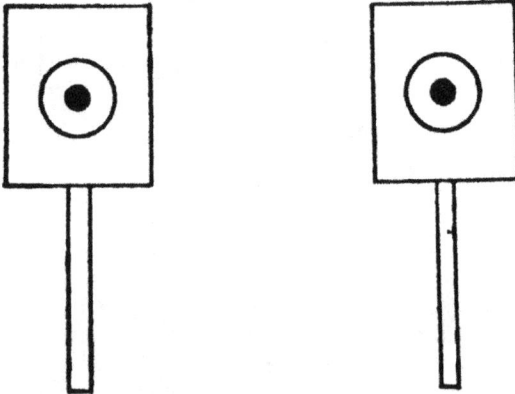

走りながら打つ

Target 4: *Nukimado* Piercing Target

The *Nukimado* is a wooden frame with a centerpiece you try to pierce with your Shuriken. The centerplate can be paper, wood, glass or metal.

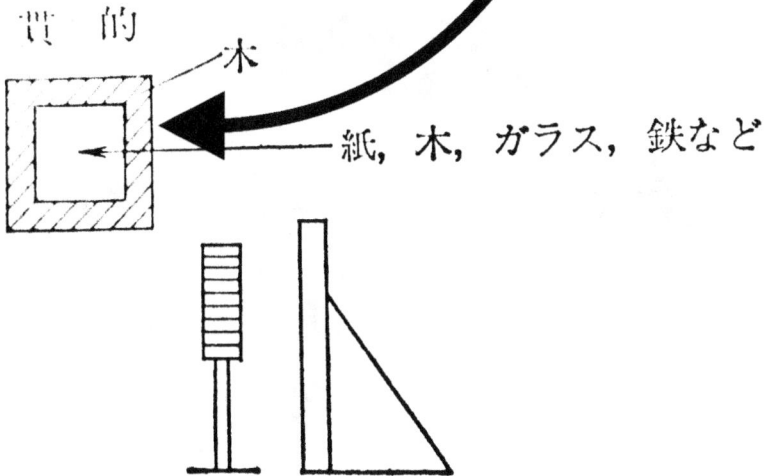

貫 的

木

紙, 木, ガラス, 鉄など

手裏剣の打ち方練習法
居打ち
How to throw Shuriken: Seated Throw

手裏剣の打ち方練習法（二）居打ち
How to Practice Throwing Shuriken: Method 2 Seated Throw
座打ち練習の標的の定め方
How to center yourself on the target for seated training.

座打ち練習の標的の定め方
心窩の上約一寸くらいのところ
と標的の中心部と一致するよう
に定めるが適当である。

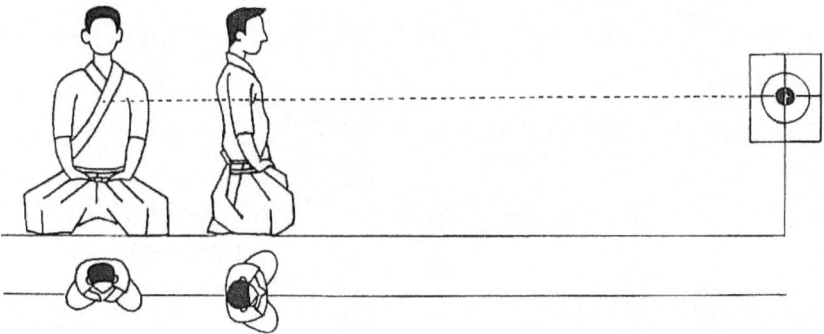

If you draw an imaginary line from the center of the target it should end approximately 1 Sun, 3 centimeters, above your heart. This is the correct height.

手裏剣の打ち方練習法（二）居打ち
How to Practice Throwing Shuriken: Method 2 Seated Throw
居打ち（座打ちともいう）練習第一の形
Throwing while seated is called E-Uchi or Za-Uchi in Japanese
Technique 1 (Steps 1-3)

手裏剣を左手に持って正面の標的に向かって正座,

標的を左にした横向きの体となり、左足を立膝とし、右足先を爪立て尻下に居敷き構える（目は終始標的から離さぬこと）

標的の中心と自己の中心をよく見定めながら、右手で左手の手裏剣をとる。

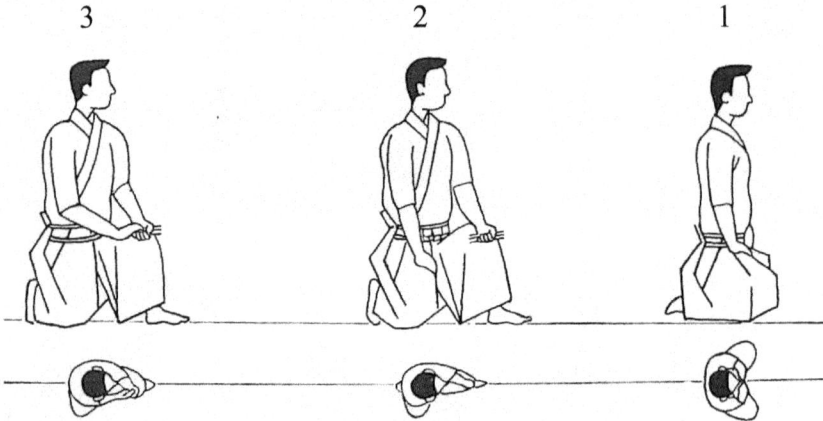

3   2   1

Take a Shuriken out of your left hand with your right hand, all the while focusing on keeping the center of the target aligned with the center of your body.	Plant your left foot in front of you and turn your left side towards the target. Your right knee is on the ground with the toes of your right foot touching the ground and your heel up. The right side of your butt should rest on your right heel. Your eyes should never divert from the target from beginning to end.	Note: Both of the Kanji 居 and 座 mean "sitting." Hold the Shuriken in your left hand and face your target in Seiza, a seating position with your legs tucked under.

手裏剣の打ち方練習法（二）居打ち
How to Practice Throwing Shuriken: Method 2 Seated Throw
居打ち（座打ちともいう）練習第一の形
Throwing while seated is called E-Uchi or Za-Uchi in Japanese
Technique 1 (Steps 4-6)

打
つ
。

右手の手裏剣を右後頭
部のやや上方に構え、

手裏剣を持った右手を
いったん右股脇に下ろ
し、もう一度標的と自
己との中心をよく定め、

| 6 | 5 | 4 |

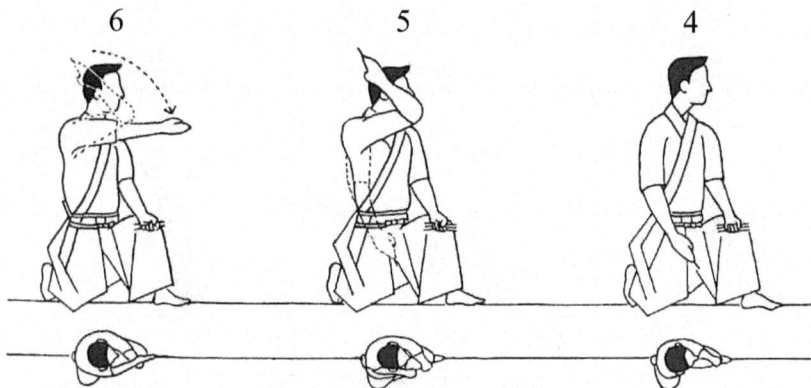

Throw	Bring your right hand behind and slightly above the right side of your head.	Lower the Shuriken in your right hand down beside your right thigh and again focus on keeping the center of your body aligned with the center of your body.

223

手裏剣の打ち方練習法（二）居打ち
How to Practice Throwing Shuriken: Method 2 Seated Throw
居打ち（座打ちともいう）練習第一の形
Throwing while seated is called E-Uchi or Za-Uchi in Japanese
Technique 1 (Steps 7-9)

打ち終ってまたもとの
正座にもどる。

上がった右手を静かに
下ろし、手裏剣を持つ
前の構えとなる。
こうして繰り返し打つ。

打った右手は打つ前の
ときのように、自然に
もとの右後頭部のわき
にかえる。

9   8   7

When you have finished return to the Seiza sitting style.	Next, lower your right hand silently so it returns to the initial position before you took a Shuriken from your left hand. Repeat these steps and throw all your Shuriken.	After throwing allow you hand to naturally return to the position behind the right side of your head, just before throwing.

手裏剣の打ち方練習法（二）居打ち
How to Practice Throwing Shuriken: Method 2 Seated Throw
居打ち（座打ちともいう）練習第一の形
Throwing while seated is called E-Uchi or Za-Uchi in Japanese
Technique 2 (Steps 1-3)

剣をとる。

ら、右手で左手の手裏

心をよく見定めなが

標的の中心と自己の中

3

標的を左にした横向き

の体となり、左足を立

膝とし、右足を中に居

敷いて構える（目は終

始標的から離さぬこと）。

2

手裏剣を左手に持って

正面の標的に向かって

正座。

1

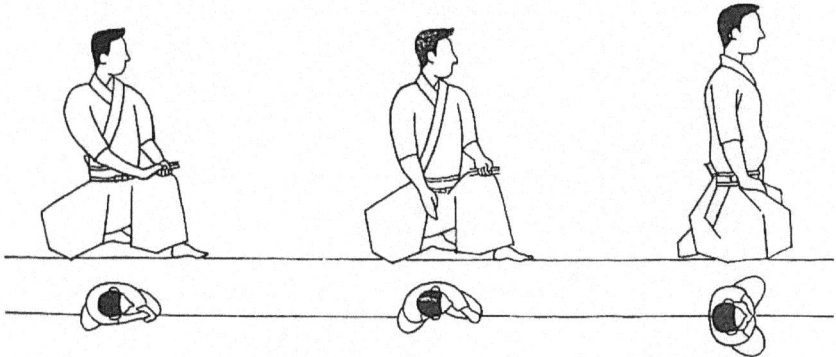

Take a Shuriken out of your left hand with your right hand, all the while focusing on keeping the center of the target aligned with the center of your body.	Plant your left foot in front of you and turn your left side towards the target. Your right knee is on the ground with the top of your right foot flat on the ground under your butt. Your eyes should never divert from the target from beginning to end.	Hold the Shuriken in your left hand and face your target in Seiza, a seating position with your legs tucked under.

手裏剣の打ち方練習法（二）居打ち
How to Practice Throwing Shuriken: Method 2 Seated Throw
居打ち（座打ちともいう）練習第一の形
Throwing while seated is called E-Uchi or Za-Uchi in Japanese
Technique 1 (Steps 4-6)

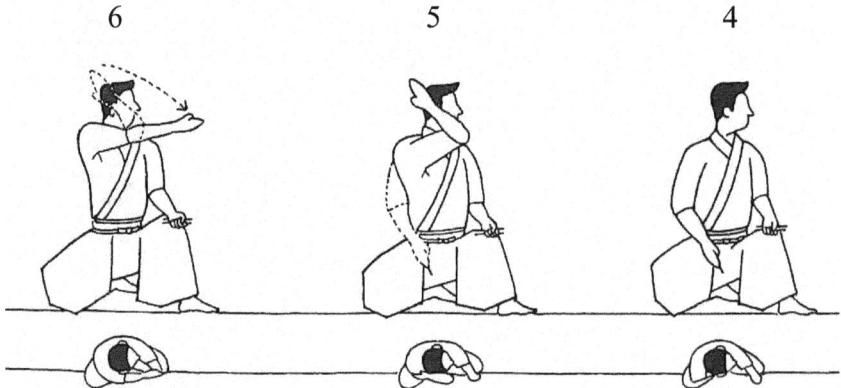

| 6 | 5 | 4 |

		手裏剣を持った右手を いったん右股脇につけ、 もう一度標的と自己と の中心をよく定め、
	右手の手裏剣を右後頭 部のやや上方に構え、	
打つ。		

Throw	Bring your right hand behind and slightly above the right side of your head.	Lower the Shuriken in your right hand down beside your right thigh and again focus on keeping the center of your body aligned with the center of your body.

手裏剣の打ち方練習法（二）居打ち
How to Practice Throwing Shuriken: Method 2 Seated Throw
居打ち（座打ちともいう）練習第一の形
Throwing while seated is called E-Uchi or Za-Uchi in Japanese
Technique 1 (Steps 4-6)

打ち終ってまたもとの正座にもどる。

上がった右手を静かに下ろし、手裏剣を持つ前の構えとなる。こうして繰り返し打つ。

打った右手は打つ前のときのように、自然にもとの右後頭部のわきにかえる。

9

8

7

| When you have finished return to the Seiza sitting style. | Next, lower your right hand silently so it returns to the initial position before you took a Shuriken from your left hand. Repeat these steps and throw all your Shuriken. | After throwing allow you hand to naturally return to the position behind the right side of your head, just before throwing. |

手裏剣の打ち方練習法（二）居打ち

How to Practice Throwing Shuriken: Method 2 Seated Throw

居打ち（座打ちともいう）練習第一の形

Throwing while seated is called E-Uchi or Za-Uchi in Japanese

居相前後打ち E-So Zengo Uchi

Throwing Forward and Back from a Seated Position
(Steps 1-3)

右後頭部のやや上方に構え、

標的の中心と自己の中心とをよく見定めながら、右手で左手の手裏剣をとり、

手裏剣を左手に持って、標的を左に、左足を立膝、右足を中に居敷いて構える。

3   2   1

Raise the Shuriken slightly above the right side of the back of your head.	Take a Shuriken from your left hand with your right hand as you ensure the center of the target is aligned with the center of your body.	With the Shuriken in your right hand, plant your left foot in front of you and turn your left side towards the target. Keep your right knee on the ground with the top of your right foot flat on the ground under your butt.

手裏剣の打ち方練習法（二）居打ち
How to Practice Throwing Shuriken: Method 2 Seated Throw
居打ち（座打ちともいう）練習第一の形
Throwing while seated is called E-Uchi or Za-Uchi in Japanese
居相前後打ち E-So Zengo Uchi
Throwing Forward and Back from a Seated Position
(Steps 4-6)

右方を横打ちに打つ。

打った右手で直ちに左手に持つ次の手裏剣をとり、

打つ。

| 6 | 5 | 4 |

Throw to your right with a Yoko Uchi, side throw.	After throwing your right hand should immediately take another Shuriken from your left hand.	Throw

229

手裏剣の打ち方練習法（二）居打ち
How to Practice Throwing Shuriken: Method 2 Seated Throw
居打ち（座打ちともいう）練習第一の形
Throwing while seated is called E-Uchi or Za-Uchi in Japanese
居打ち練習 E-Uchi Renshu
Seated Throwing Practice

聞いて打つ。

そのまま打つ。

座っているとき。

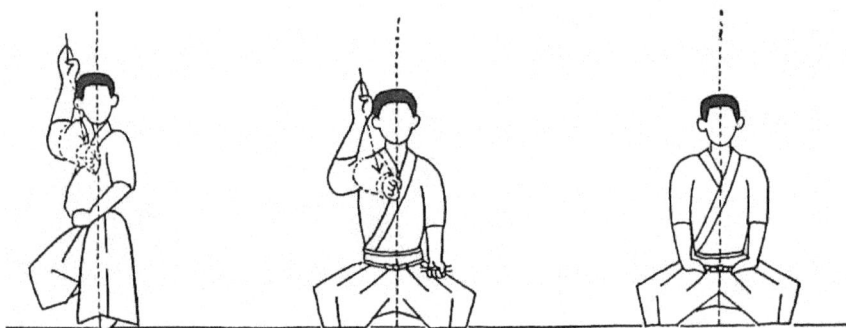

Stepping forward and throwing.	Throwing directly from a seated position	Seated position

手裏剣の打ち方練習法（二）居打ち
How to Practice Throwing Shuriken: Method 2 Seated Throw
居打ち（座打ちともいう）練習第一の形
Throwing while seated is called E-Uchi or Za-Uchi in Japanese
居打ちによる本打ち、横打ち、逆打ちの練習
Practice throwing Main Throw, Side Throw and Reverse Throw From a Seated Position

逆打ち　　　　　横打ち　　　　　本打ち

Gyaku Uchi Reverse Throw	Yoko Uchi Side Throw	Hon Uchi Main Throw

早打ち
Haya Uchi : Rapid Throw
一気五剣
Ikki Go Ken : Five Blades in Rapid Succession

早打ち（一気五剣）

手裏剣の早打ちというのは、最初の手裏剣を打って、その手裏剣が未だ標的に達しないうちに、つぎつぎと連続的に次の手裏剣を打つ早打業のことで、敵に前進どころか、立ち直る隙も与えぬような打ち方をする方法である。

これには「一気五剣」といって、一呼吸の間に五本は打てるだけの練習をすべきである。

Shuriken Haya Uchi means after you throw your first Shuriken to throw the following Shuriken one after the other rapidly enough so that the last Shuriken leaves your hand before the first one strikes the target. This is a Haya Waza, Speed Technique, used to stop an advancing enemy or to use in order to create a chance to escape.

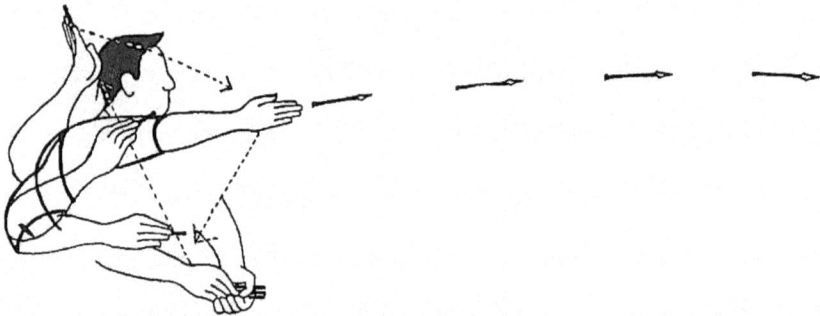

Ikki Go Ken, Five Blades in Rapid Succession, is something that should be trained so that you can throw all five Shuriken in the span of one breath.

刀法
併用
手裏剣術

Using the Katana With Shuriken

刀法供用手裏剣術 *Toho Kyoyo Shuriken Jutsu*
Using the Katana With Shuriken

三学
敵と我れとの間積りは
五寸三間と定め　先づ
身體を実直にして（檣
造）討つ　我が天眞と
り打出す時　息を詰め
て打掛ける　是れを躰
と手と息の三學といふ

剣当の事
手裏剣で主に打ち當て
る所を剣當と云ひ三ヶ
所ある　両眼（是れを
二星の當りと云ひ）
息下　胸中（是れを息
の當りと云ふ）両眼
息下　胸中で三ヶ所也

釼込之大事
手離を以って敵の躰を
切る　これを釼込と云
ふ　両眼の真中　鼻中
の真中　息下の真中や
咽脈の通りを打つ十の
・と云ふ習がある

三学
Sangaku : Three Lessons

The rule is to advance on your enemy in 5 Sun, 15 centimeter, intervals. Hold your body strictly upright and throw. To throw naturally you need to stop breathing and throw. The Three Lessons are the Body, the Hand and Breathing.

剣当の事
Ken Atari no Koto : Where Your Blade Should Strike

There are three primary places you should aim for with your Shuriken.
Ryogan : Both eyes. This is also known as Nisei no Atari, Striking the Two Stars
Iki no Shita : Whispering
Kyochu : Center of the Chest/ One's Intention
Both Eyes, Whispering and Center of the Chest are the three points.

釼込み之大事
Tsurugi Komi no Daiji : Drawing Blood With Your Blade

If the Shuriken you throw cuts your enemy then it is called Drawing Blood With Your Blade.
For example striking directly in the eye, in the nose or in a vein. This is known as The Lesson of Ten [illegible.]

刀法供用手裏剣術 *Toho Kyoyo Shuriken Jutsu*
Using the Katana With Shuriken

三學

敵と我れとの間積りは
五寸三間と定め　先づ
身體を實直にして（體
違）討つ　我が天窟よ
り打出す時　息を詰め
て打掛ける　是れを躰
と手と息の三學といふ

観当の事

手裏剣で主に打ち當て
る所を剣當と云ひ三ケ
所ある　両眼（是れを
二星の当りと云ひ）
息下　胸中（是れを息
の当りと云ふ）　両眼
息下　胸中で三ケ所也

観込之大事

手裏を以って敵の躰を
切る　これを観込と云
ふ　両眼の真中　鼻中
の真中　息下の真中等
狂脈の通りを打つ十の
と云ふ習がある

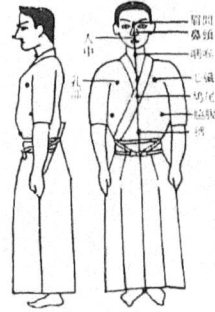

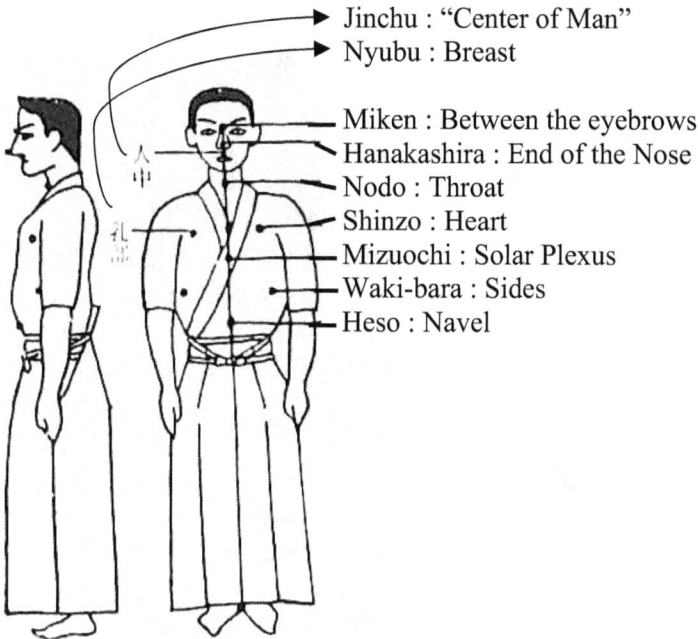

Jinchu : "Center of Man"
Nyubu : Breast

Miken : Between the eyebrows
Hanakashira : End of the Nose
Nodo : Throat
Shinzo : Heart
Mizuochi : Solar Plexus
Waki-bara : Sides
Heso : Navel

刀法供用手裏剣術 *Toho Kyoyo Shuriken Jutsu*
Using the Katana With Shuriken
実戦に臨むときの心得
Lessons on How to Fight

Kage Ken

Kage Ken

Shuriken

実戦に臨むときの心得

実戦に臨むとき、手裏剣は必ず、左側腹脇前に数本差し添えて持ったものである。甲賀流では手裏剣のほか、剣形手裏剣を差し添える習わしであった。また、車剣は細い鉄棒に差し持ったものである。映画演戯等でよく鉢巻のところに、手裏剣をかんざしのように差したりするのを見るが、あんなことは絶対になかったことである。敵に我が手裏剣の持数を知られることは非常に不利であるため、心得ある武士はしなかった。また陰剣といって、剣形手裏剣（普通の手裏剣でも）を右側後腰脇か右えり下懐中に隠す習いもある。

When faced with going into battle you should always have several Shuriken secured in your belt on your left side. The Koga School also instructs its students to carry Katana Gata, Sword Shaped, Shuriken in addition to regular Shuriken. The Shaken, multi-pointed Wheel Shuriken, should be threaded through with a thin piece of wire. In the movies you often see hairpin shaped Shuriken stuck in the Hachimaki headbands worn by people, however this is absolutely unrealistic.

It is important to never allow the enemy to see how many Shuriken you carry, as this will put any Samurai at a disadvantage. Kage Ken, Shadow Blades, are the solution to this. Keep several Sword Shaped Shuriken or regular Shuriken, tucked either on your back on the right hand side or inside the base of your shirt.

中刺

車剣 Kuruma Ken
"Wheel" Shuriken

刀法供用手裏剣術 *Toho Kyoyo Shuriken Jutsu*
Using the Katana With Shuriken
How to Fight Technique 1 (Steps 1-3)

左足を後方に引きながらすばやく刀を抜き、

3

右手を柄にかけ、

2

敵と相対し、

1

3	2	1
Rapidly draw your Katana as you pull your left foot back.	Place your right hand on the handle of your Katana.	Face your opponent.

刀法供用手裏剣術 *Toho Kyoyo Shuriken Jutsu*
Using the Katana With Shuriken
How to Fight Technique 1 (Steps 4-6)

6

右側後頭部上に
打ち構えに構え
る。

5

右手より左手に
刀の柄を持ちか
えると同時に、
右手は左腹脇下
に隠くし持った
手裏剣をとり、
右足を一歩後へ
引きながら、

4

左手を添えて晴
眼に構えると見
せ、

6	5	4
…raise the Shuriken up above the right side of the back of your head. This is the ready position.	As soon as your left hand takes hold of the handle of your sword, draw a Shuriken from where you have hidden them on the left side of your waist. As you draw your right foot back…	Join your left hand to your right and stare hotly at your opponent.

Translator's note : Fujita Seiko uses the unusual Kanji combination of 暑眼 for "stare hotly." The Kanji 暑 is a variant of 暑, which means "summer heat" and is not used in Japanese.

刀法供用手裏剣術 *Toho Kyoyo Shuriken Jutsu*
Using the Katana With Shuriken
How to Fight Technique 1 (Variations A ~ C)

この打ち構えの
場合、刀を晴眼
に構えるのと（一）
突き出して構え
るのと（二）
垂直に構える（三）
のとがあるが、
そのときの場合
によっていずれ
にても可し。

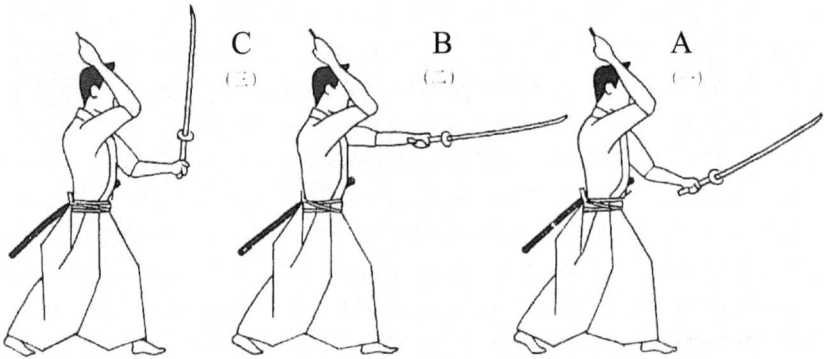

C (三)　　B (二)　　A (一)

C	B	A

From this position there are three options:
A: Keep the Katana in Seigan, with the tip aimed at the attacker's eyes.
B: Hold your Katana straight out, perpendicular to the ground.
C: Hold your Katana vertically.

刀法供用手裏剣術 *Toho Kyoyo Shuriken Jutsu*
Using the Katana With Shuriken
How to Fight Technique 1 (Steps 7-9)

9	8	7
The moment you finish throwing your second Shuriken....	Immediately grab another Shuriken.	Throw the Shuriken you readied.

刀法供用手裏剣術 *Toho Kyoyo Shuriken Jutsu*
Using the Katana With Shuriken
How to Fight Technique 1 (Steps 10-12)

敵のようすを見ながら、

踏み込んで切り、

刀を上段に振りかぶり、

12

11

10

12	11	10
As you maintain a close watch on your enemy…	Cut down as you step forward.	…raise your Katana over your head. This is Jodan Kamae, Upper Stance.

刀法供用手裏剣術 *Toho Kyoyo Shuriken Jutsu*
Using the Katana With Shuriken
How to Fight Technique 1 (Steps 13-14)

終る。

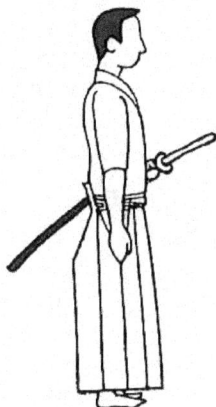

静かに納刀し、

14

13

14	13
End of this technique.	…sheathe your Katana.

刀法供用手裏剣術 *Toho Kyoyo Shuriken Jutsu*
Using the Katana With Shuriken
How to Fight Technique 2 (Steps 1-3)

打つ。

打ち構えに横えながら敵にせまり、

敵と相対するや直ちに手裏剣を取り、

3　　　　　　　　**2**　　　　　　　　**1**

3	2	1
…and throw.	As you move into your throwing stance, advance on the enemy…	As soon as you encounter an enemy, reach for a Shuriken.

刀法供用手裏剣術 *Toho Kyoyo Shuriken Jutsu*
Using the Katana With Shuriken
How to Fight Technique 2 (Steps 4-6)

振りかぶって、

刀を抜いて、

打った手は直ちに柄にかけ、

6

5

4

6	5	4
Raise your Katana up and…	Draw your Katana.	After releasing the Shuriken, that hand moves to grasp the handle of your Katana.

刀法供用手裏剣術 *Toho Kyoyo Shuriken Jutsu*
Using the Katana With Shuriken
How to Fight Technique 2

切る。

7
...cut.

刀法供用手裏剣術 *Toho Kyoyo Shuriken Jutsu*
Using the Katana With Shuriken
How to Fight Technique 3

<table>
<tr>
<td>実戦に臨むときの心得第三 敵を左右に受けた場合。</td>
<td>

実戦に臨むときの心得第三
Lessons on How to Fight Technique 3
How to defend against two opponents, one on your left and one on your right.

</td>
</tr>
</table>

3	2	1
手裏剣を取るやいなや、	左右の敵を警戒しながら、まず左方の敵に対す、	左手を鞘の鯉口にかけ、
Draw and throw a Shuriken…	While being alert to the opponents to your left and right, ready yourself to attack left first.	Grip the scabbard of your sword at the point it meets the sword guard. This point is called the Koiguchi "carp's mouth."

刀法供用手裏剣術 *Toho Kyoyo Shuriken Jutsu*
Using the Katana With Shuriken
How to Fight Technique 3

振りかぶって、

6

抜刀して、

5

柄に手をかけ、

4

6	5	4
Raise it up.	Draw your Katana.	…then immediately grab the handle of your Katana.

刀法供用手裏剣術 *Toho Kyoyo Shuriken Jutsu*
Using the Katana With Shuriken
How to Fight Technique 3

振りかぶって　9

右方え向きかえりながら、　8

切り。　7

9	8	7
Bring your Katana up…	Shift to facing the opponent to your right.	…and cut.

刀法供用手裏剣術 *Toho Kyoyo Shuriken Jutsu*
Using the Katana With Shuriken
How to Fight Technique 3

切る。

10
…and cut.

刀法供用手裏剣術 *Toho Kyoyo Shuriken Jutsu*
Using the Katana With Shuriken
How to Fight Technique 4

<table>
<tr><td>

実戦に臨むときの第四
右方に敵を受け
た場合。

</td><td>

実戦に臨むときの心得第四
Lessons on How to Fight Technique 4
How to defend against an opponent to your right.

</td></tr>
</table>

逆打ち、

手裏剣を取り、

| 3 | 2 | 1 |

3	2	1
Do a Gyaku Uchi, Reverse Throw.	Grab a Shuriken.	[no text]

251

刀法供用手裏剣術 *Toho Kyoyo Shuriken Jutsu*
Using the Katana With Shuriken
How to Fight Technique 4

6	5	4
Raise your Katana.	Draw your Katana.	Grab the handle of your Katana.

刀法供用手裏剣術 *Toho Kyoyo Shuriken Jutsu*
Using the Katana With Shuriken
How to Fight Technique 4

切る。

7
Cut.

刀法供用手裏剣術 *Toho Kyoyo Shuriken Jutsu*
Using the Katana With Shuriken
How to Fight : Secret Teachings

すべて敵と対す
るときは、常に
左の腹脇下前に
手裏剣を五本持
つことはもちろ
んであるが、そ
の手裏剣のほか
に陰剣を持つこ
とが秘伝である。

Obviously you should carry five Shuriken on your left side, tucked into your belt to use against opponents. However you should also carry In-ken, Concealed Shuriken. This is a Hiden, or Secret Teaching.

Translator's Note :
The five Shuriken you carry on your left side are the Yo-ken, or Yang Shuriken, meaning they are the observable Shuriken. The In-Ken are the Yang Shuriken, which are hidden. The two illustrations show the Samurai taking Shuriken from the two In-Ken storage points.

刀法供用手裏剣術 *Toho Kyoyo Shuriken Jutsu*
Using the Katana With Shuriken
知新流
Chishin School

手裏剣を右手に隠し持って敵と相対し、

敵、刀の柄に手を掛けると見るや、右足を敵の目を目当てに踏み出すとともに手裏剣を打ち、手は直ちに柄にかけ、

Chishin School Fundamental

Conceal the Shuriken in your right hand as you face the enemy.

As soon as you see your enemy move his hand to the Tsuka, or handle, of his sword, step forward with your right foot. Throw the Shuriken, aiming for his eyes, then immediately grab the Tsuka of your own Katana.

敵がひるむとこ
ろを

踏み込んで切る。

257

Chishin School Fundamental

As your enemy is reacting to your strike…

…step in and cut.

立礼して後、

一本目

手裏剣五本を右前半に差し、日本刀を右手に提げて出る。

Chishin School Technique 1

1	2
Start with five straight Shuriken tucked into the front of your Obi. Your Nihonto, Katana, is in your right hand as you step forward and stop.	Do a Tachi-rei, or standing bow, then…

まず左手で鞘口を持ち、鍔五分ばかり離して刀の柄を握り、右足を斜め前方に踏み出し、

刀を左手に持ちかえて帯刀する。

Chishin School Technique 1	
3	**4**
Slide your Katana into your Obi and hold onto the scabbard with your left hand.	Your first move is to open the Koiguchi. This means to push on the Tsuba, sword guard, with your left thumb, open 1.5 centimeters, unsticking the sword from it's scabbard. Grip the Tsuka with your right hand and step diagonally forward with your right foot.

右足をもどし、柄と右手腕とを併行させ、右胸脇に構え、左手で鞘口を握ったまま直立して前方標的に見入る。

刀を抜き、片手八相に構え、

Chishin School Technique 1	
5	**6**
Draw your Katana and hold it in your right hand in Hasso Kamae, with the blade vertical. Then step back with your right foot to your starting stance.	As you step back, ensure the handle of your sword as well as your right hand and arm all move in unison with your body. Stay in this Migi Waki Gamae, or Stance With the Sword by Your Right Side. Keeping your left hand on the Koiguchi, stand straight and stare directly at the person who is your target.

持ち終ったら、手を下げ、標的に向かって散歩進む。

次に刀を左手に持ちかえて構え、前方を凝視しつつ、右手で手裏剣を一本抜き出し右手に持つ。

Chishin School Technique 1

7	8
As you stare intensely at the enemy in front of you, switch the Katana to your left hand. With your right hand remove one Shuriken from your belt and hold it in your right hand.	After you have positioned the Shuriken in your hand allow your arm to hang down. Take several steps towards your target.

左上段に構えた刀をさらに高く構え、手裏剣を打つと、直ちに、

間合をはかって気合とともに左手に持った刀を左片手上段に構え、手裏剣を打ち構えにして、左半身になってじりじりと敵にせまる。

9	10
Focus all your energy as you judge the distance to your target. Bring your left hand holding the Katana up to Katate Jodan, One-handed upper stance and, at the same time, ready your right hand to throw the Shuriken. Keep your left side forward as you slowly and steadily advance on your target.	Continue to raise your Katana to a higher Jodan Kamae and throw your Shuriken. Then immediately...

刀を八相に構える。

11

...go into Haso Kamae, with the Katana by your side.

次に刀の切先を前
方に向け、左足を
斜め左後方に引き
右手の刀は前方に
突き出し、左手は
すばやく鞘口を握
り、納刀の姿勢に

12

Next, point the end of your Katana forward and pull your left foot diagonally backwards. Hold the Katana in your right hand and push your arm forward. Your left hand should go to the Koiguchi, carp's mouth, of your scabbard in a quick motion. Begin sheathing your sword.

左手は刀の鞘口を
握り、左足をもど
し、もとの自然体
となり、一本目終
る。

しずかに納刀。

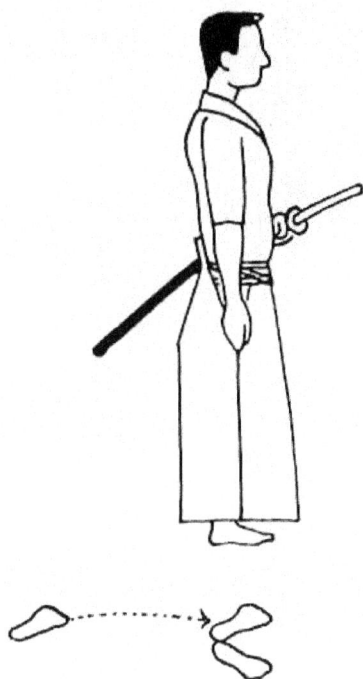

Chishin School Technique 1	
13	**14**
Quietly sheathe your Katana.	With your left hand still on the Koi-guchi, move your left foot beside your right and return to your natural standing position.
	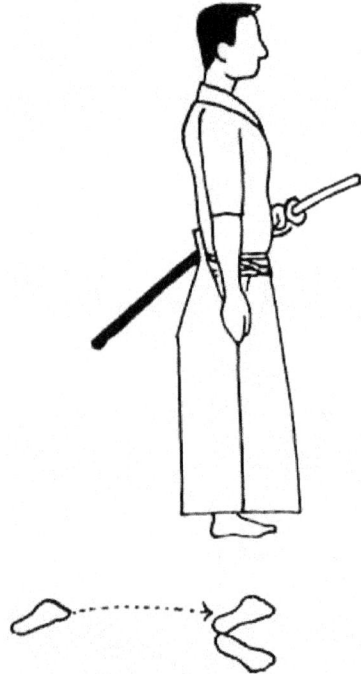
Note: The following page shows the entire sequence	

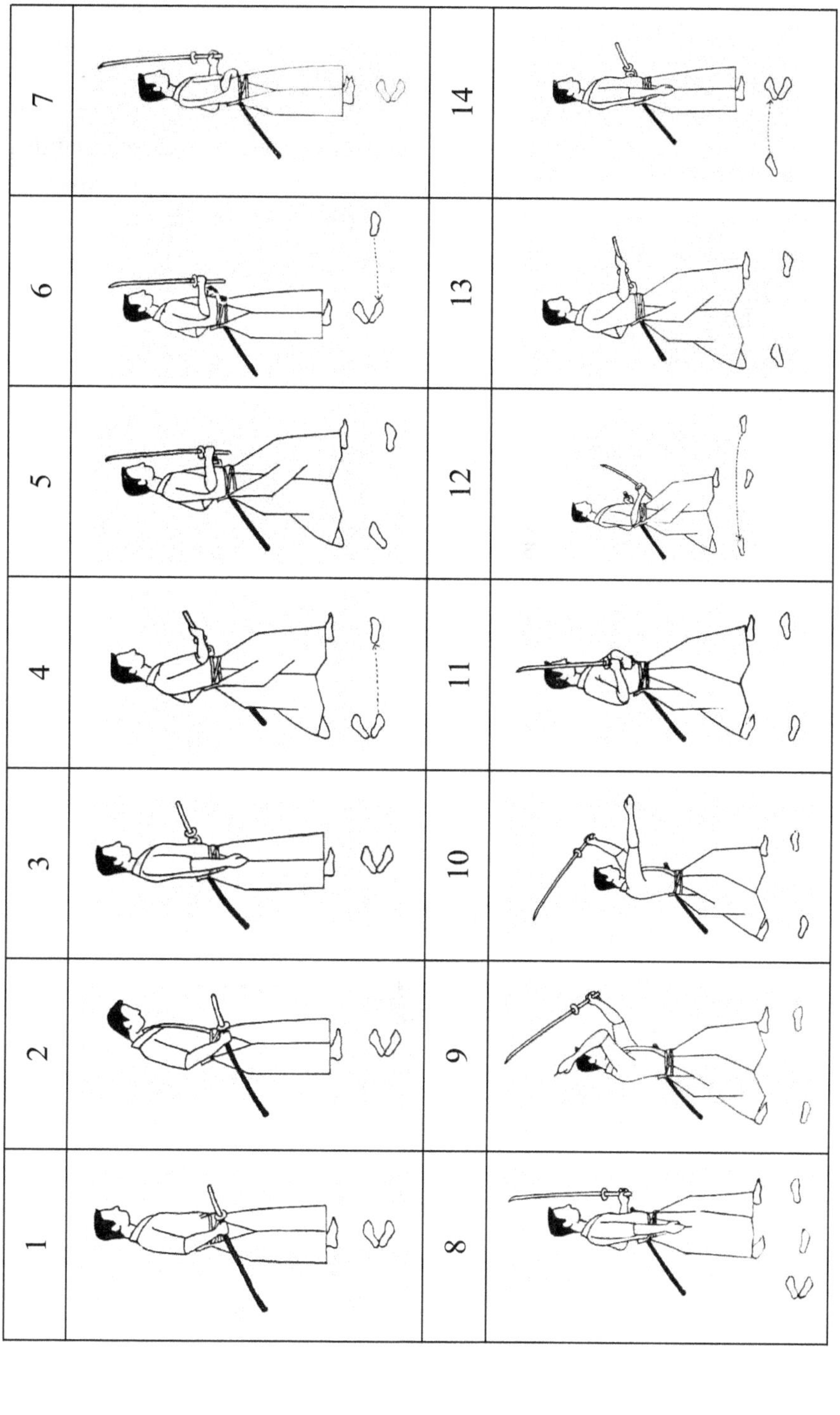
1
2
3
4
5
6
7
8
9
10
11
12
13
14

Chishin School Technique 2

Hold your Katana straight up in front of you with your left hand. From that stance throw your Shuriken. Use the Tsuba, sword guard, to guide your throw. Keep the edge of the Tsuba on his face and aim to throw between his eyes.

二本目

左手に持った刀を
垂直に前にさし出
して構えながら打
つ。このとき、刀
の鍔のふちを敵の
顔面、両眼の間に
つけて目標とする。

Chishin School Technique 3

Holding the Katana in your left hand with your arm extended, straight out, throw your Shuriken. The tip of your sword is forward. When doing this technique keep the tip of the sword on the enemy's face, directly between his eyes.

三本目
刀を左下に持ち、
左直向に切先を突
き出しながら打つ。
このとき刀の切先
は、敵の顔、両眼
の間につける。

同じく四本目の替手
柄頭を左に切先を
右に向けて持って
打つ。

四本目
刀を左下に持ち、
切先を前方に向け
て下げながら打つ。

Chishin School Technique 4

Throw while holding your Katana downward in your left hand with the point forward.	This is a different way to do Technique 4. The Tsuka-gashira, or pommel, is facing to your left and the Kisaki, tip of the sword is facing to your right as you throw.

打ち終ったら、す
ばやく刀を抜いて
構え、

五本目
刀を抜かず、柄を
斜前方に構えて打
つ。

Chishin School Technique 5

Do not draw your sword but push the Tsuka, or handle, diagonally downward. Throw from this position.

After you throw your Shuriken rapidly draw your sword and stand as shown in the illustration.

納刀して終る。

次に切先を前方に
向け突き出して後、

Chishin School Technique 5

Next, stab forward with the tip of your sword, then...	...sheathe your sword and the technique is over.

Overview of steps 1 ~4 of Technique 5.

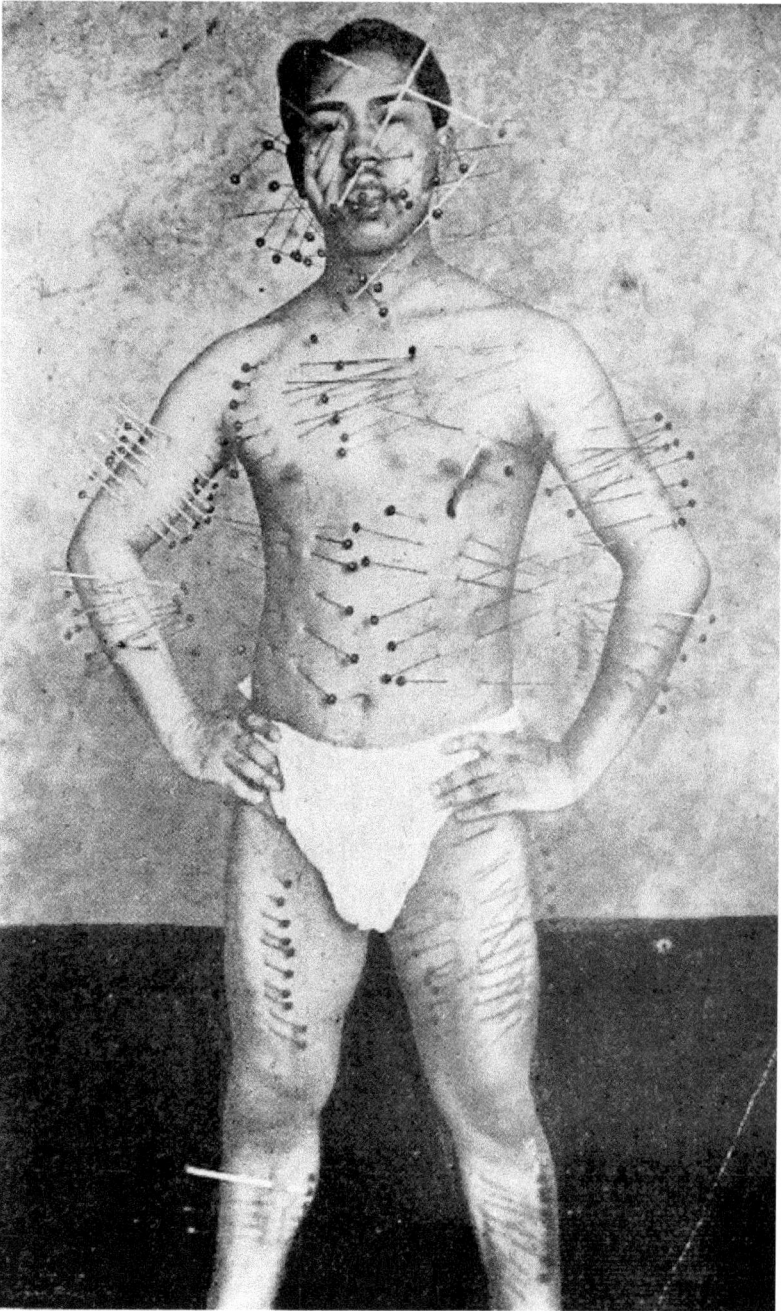

Picture of Fujita Seiko from his 1936 book *Secrets of Ninjutsu* pierced with 258 tatami needles.

About the Author:

Fujita Seiko 藤田 西湖
1898 –1966

Fujita Seiko was a prominent martial artist, researcher and author. He was born in Tokyo, and studied Kōga-ryū Ninjutsu under the tutelage of his grandfather the 13th head of the tradition. He later became the 14th and final head of that school. Throughout his life Fujita Seiko continued to study and research martial arts and published numerous books on all aspects of Japanese Budo.

Works by Fujita Seiko translated by Eric Shahan:

What is Ninjutsu? 1938
The 18 Weapons of War 1958
Samurai Bondage Volume 1 1964
An Illustrated Guide to Shuriken 1964

www.ingramcontent.com/pod-product-compliance
Lightning Source LLC
Chambersburg PA
CBHW060333200326
41519CB00011BA/1922